CONTINUOUS PSEUDOMETRICS

LECTURE NOTES IN PURE AND APPLIED MATHEMATICS

1. *N. Jacobson,* Exceptional Lie Algebras
2. *L.-Å. Lindahl* and *F. Poulsen,* Thin Sets in Harmonic Analysis
3. *I. Satake,* Classification Theory of Semi-Simple Algebraic Groups
4. *F. Hirzebruch, W. D. Neumann,* and *S. S. Koh,* Differentiable Manifolds and Quadratic Forms
5. *I. Chavel,* Riemannian Symmetric Spaces of Rank One
6. *R. B. Burckel,* Characterization of C(X) among Its Subalgebras
7. *B. R. McDonald, A. R. Magid,* and *K. C. Smith,* Ring Theory: Proceedings of the Oklahoma Conference
8. *Yum-Tong Siu,* Techniques of Extension of Analytic Objects
9. *S. R. Caradus, W. E. Pfaffenberger, and Bertram Yood,* Calkin Algebras and Algebras of Operators on Banach Spaces
10. *Emilio O. Roxin, Pan-Tai Liu,* and *Robert L. Sternberg,* Differential Games and Control Theory
11. *Morris Orzech* and *Charles Small,* The Brauer Group of Commutative Rings
12. *S. Thomeier,* Topology and its Applications
13. *Jorge M. López* and *Kenneth A. Ross,* Sidon Sets
14. *W. W. Comfort* and *S. Negrepontis,* Continuous Pseudometrics

Other volumes in preparation

CONTINUOUS PSEUDOMETRICS

W. W. COMFORT
Wesleyan University
Middletown, Connecticut

S. NEGREPONTIS
Athens University
Athens, Greece
and
McGill University
Montréal, Canada

MARCEL DEKKER, INC. New York

MARCEL DEKKER, INC.

270 Madison Avenue, New York, New York 10016

LIBRARY OF CONGRESS CATALOG CARD NUMBER: 75-13800

ISBN: 0-8247-6294-0

Current printing (last digit):
10 9 8 7 6 5 4 3 2 1

PRINTED IN THE UNITED STATES OF AMERICA

INTRODUCTION

While collaborating in research in the summers of 1966 and 1967, we developed a point of view concerning a body of interesting topological theorems which seemed to us different from, and more natural than, any view we had seen systematically exploited before. Experiences teaching at McGill University and at Wesleyan University appeared to confirm the utility of our observation. These Notes, offered now at the suggestion of friends and former students, are essentially the classroom notes we developed during the academic year 1967-68.

Among topological spaces, metrizable spaces have many desirable and special properties. Given a space X and a continuous function f onto a metrizable space M, it is often possible to use f^{-1} to "pull back" a particular property from M to X. In order that M and f can be defined, it is enough to have a continuous pseudometric defined on X: one then defines M to be the metric space of equivalence classes and f the canonical quotient mapping.

Continuous pseudometrics are constructed here in two simple ways: from a countable or a locally finite family of cozero-sets (of continuous functions into $[0,1]$). For example if $\{U_i : i \in I\}$ is such a locally finite family, with $U_i = \operatorname{coz} f_i$, then $\rho(x,y) = \Sigma_{i\in I} |f_i(x) - f_i(y)|$ is a continuous pseudometric; and if $\{U_n : n < \omega\}$ is such a countable family, with $U_n = \operatorname{coz} f_n$, then $\rho(x,y) = \Sigma_{n<\omega} |f_n(x) - f_n(y)|/2^n$ is a continuous pseudometric.

Among the applications presented here of this technique are these: Frolík's theorem (9.4) that a paracompact space X is a G_δ in its Stone-Čech compactification βX if and only if there is a perfect function from X onto a complete metric space; the theorem of Frolík and Negrepontis (9.6) that a space X is a Baire set in βX if and only if there is a perfect function from X onto a separable metrizable absolute Borel space; the theorem of Lorch and Negrepontis (9.10) that a Baire set in a realcompact [resp., topologically complete] space is realcompact [resp., topologically complete]; the theorem of Kodaira-Halmos-Comfort (9.15) that if X is a Baire set in βX then every closed Baire set of X is a zero-set; Tamano's theorem (5.3) that X is paracompact if and only if $X \times \beta X$ is a normal space; and the Katětov-Shirota theorem (6.3) that a topologically complete space is realcompact if and only if it contains no closed, discrete subset of Ulam-measurable cardinal.

W. Wistar Comfort
Stylianos Negrepontis

Contents

CONTINUOUS PSEUDOMETRICS

§1. $\bar{P}$-REFLECTIONS

We begin with the basic definitions and notation to be used in these Notes.

By *space* we mean a completely regular, Hausdorff topological space. Thus for every space X we have

(i) if $p, q \in X$ and $p \neq q$ then there are disjoint open subsets U and V of X such that $p \in U$ and $q \in V$; and

(ii) if F is a closed subset of X and $p \in X \backslash F$, then there is a continuous function

$$f : X \to [0,1]$$

such that $f(p) = 0$ and $f(x) = 1$ for $x \in F$.

Definition. Let X be a set (or a space). Then id_X is the function from X to X defined by $id_X(x) = x$.

Notation. $|X|$ = cardinality of X.

$P(X) = \{A : A \subset X\}$.

$c\ell_X A$ denotes the closure in X of A.

If $f : X \to Y$ and $A \subset X$, then $f|A$ denotes the restriction of f to A. That is,

$$f|A = \{\langle x,y\rangle \in X \times Y : x \in A \text{ and } y = f(x)\}.$$

If X and X' are homeomorphic spaces we write $X \cong X'$.

Definitions. Let P be a class of spaces and let X and Y be spaces with $X \subset Y$.

(a) X is *P-embedded* in Y if for every continuous function $f : X \to Z \in P$ there is a continuous function $\tilde{f} : Y \to Z$ such that $f \subset \tilde{f}$;

(b) Y is a *P-fication* of X if $Y \in P$ and X is a dense subspace of Y; and

(c) Y is a *P-reflection* of X (or, a *reflection of X in P*) if Y is a P-fication of X and X is P-embedded in Y.

A class P of spaces is said to be

(a) *closed under homeomorphisms* (or, *$\cong$-closed*) if $X' \cong X \in P \Rightarrow X' \in P$;

(b) *finitely productive* if $X, Y \in P \Rightarrow X \times Y \in P$;

(c) *productive* if $X_i \in P$ for $i \in I \Rightarrow \prod_{i \in I} X_i \in P$;

(d) *closed-hereditary* if $X \in P$, A closed in $X \Rightarrow A \in P$.

It is convenient to consider the smallest $\cong$-closed, productive, closed-hereditary class containing a given class P.

Notation. Let P be a non-empty class of spaces. Then $\bar{P}$ denotes the class of all spaces homeomorphic to a closed subspace of a product of elements of P.

In category-theoretic terminology, $\bar{P}$ is called the *epi-reflective hull of* P (in the class of completely regular Hausdorff spaces).

We note that $\bar{\bar{P}} = \bar{P}$ for every class P.

We denote by C the class of compact spaces. A C-fication of a space X is called a *compactification* of X.

1.1. *Lemma*. Let Z be a space and X dense in Z.

Then $|Z| \le 2^{2^{|X|}}$.

Proof. For $z \in Z$ set

$$\mathcal{U}(z) = \{U \cap X : U \text{ is a neighborhood of } z\};$$

we have $\mathcal{U}(z) \subset P(X)$ and hence $\mathcal{U}(z) \in P(P(X))$. Further the function $z \to \mathcal{U}(z)$ is one-to-one, for if z_0 and z_1 are distinct elements of Z and U is a neighborhood of z_0 such that $z_1 \notin cl_Z U$, then

$$U \cap X \in \mathcal{U}(z_0) \text{ and } U \cap X \notin \mathcal{U}(z_1).$$

1.2. *Theorem.* Let P be a class of spaces and let X be a space with a $\bar{P}$-fication Y. Then X has a $\bar{P}$-reflection.

Proof. If $|X| < \omega$ we have $X = Y \in \bar{P}$ (since X is dense in Y) and hence X is itself a $\bar{P}$-reflection of X. We assume then that $|X| \ge \omega$.

Let $\{<Z_i, f_i> : i \in I\}$ be a list of all spaces $Z_i \in \bar{P}$ and continuous functions $f_i : X \to Z_i$ such that

$$Z_i \subset P(P(X)) \text{ and } f[X] \text{ is dense in } Z_i.$$

Let j denote the inclusion function from X into Y. Since $|Y| \le |P(P(X))|$ by Lemma 1.1, there are $i \in I$ and a homeomorphism h of Y onto Z_i such that $f_i = h \circ j$; in particular $I \ne \emptyset$. We note further that the number of functions from X into $P(P(X))$ does not exceed $(2^{2^{|X|}})^{|X|}$, the number of subsets of $P(P(X))$ is $2^{2^{2^{|X|}}}$, and the number of topologies on a set Z does not exceed $2^{2^{|Z|}}$; hence I is a set. We define $Z = \prod_{i \in I} Z_i$, we define $e : X \to Z$ by

$$e(x)_i = f_i(x),$$

and we prove that e is a homeomorphism.

e *is continuous.* For $i \in I$ the function $\pi_i \circ e$ is just f_i. Hence e is continuous.

e *is one-to-one.* Let i, h and j be as above, so

that $f_i = h \circ j$. Then for distinct elements x_0 and x_1 of X we have

$$e(x_0)_i = f_i(x_0) = h(x_0) \neq h(x_1) = f_i(x_1) = e(x_1)_i.$$

e *is open (onto* e[X]). Let U be an open subspace of X, let i and h be as above, and let V be an open subset of Y such that $U = V \cap X$. Then

$$e[U] = e[X] \cap \pi_i^{-1}(h[V]),$$

so that e[U] is (relatively) open in e[X]. The proof that e is a homeomorphism is complete.

Finally we set $\bar{X} = c\ell_Z\, e[X]$. Since $Z_i \in \bar{P}$ for $i \in I$ we have $\bar{X} \in \bar{\bar{P}} = \bar{P}$ and hence $\bar{X}$ is a $\bar{P}$-fication of e[X]. To see that $\bar{X}$ is a $\bar{P}$-reflection of e[X] let f be a continuous function from e[X] into a space $K \in \bar{P}$. Replacing K if necessary by $c\ell_K\, f[e[X]]$ we assume that f[e[X]] is dense in K, so that $|K| \leq 2^{2^{|X|}}$ by Lemma 1.1; thus we may also assume that $K \subset P(P(X))$. There is $i \in I$ such that $\langle K, f \circ e\rangle = \langle Z_i, f_i\rangle$, and for $x \in X$ we have

$$f(e(x)) = f_i(x) = \pi_i(e(x));$$

hence $\pi_i : Z \to Z_i$ is a continuous extension of $f \circ e$ and we have

$$f \subset \pi_i | \bar{X} : \bar{X} \to K.$$

Thus $\bar{X}$ is a $\bar{P}$-reflection of e[X]. It follows that X has a $\bar{P}$-reflection. The proof is complete.

1.3. *Corollary* (Tychonoff; Stone-Čech). Every space has a C-reflection.

Proof. By the Tychonoff product theorem we have $C = \bar{C}$. By another theorem of Tychonoff every space has a C-fication. (The proof is similar to the proof of Theorem 1.2. Let X be a space, let F be the family of all continuous functions $f : X \to [0,1]$, and define $e : X \to \prod_{f \in F} [0,1] = P$ by the rule $e(x)_f = f(x)$. Then e is a homeomorphism, and $c\ell_P\, e[X]$ is a C-fication of e[X]). The result now follows from Theorem 1.2.

1.4. *Lemma.* Let $f, g : X \to Z$ be continuous, with Z

a Hausdorff space. Then

$$A = \{x \in X : f(x) = g(x)\}$$

is closed in X.

Proof. Let $x \in X$, $f(x) \neq g(x)$, and let U and V be disjoint neighborhoods in Z of f(x) and g(x). Then

$$f^{-1}(U) \cap g^{-1}(V)$$

is a neighborhood of X disjoint from A.

1.5. *Lemma.* Let $f : Y \to Z$ be continuous, X dense in Y and $f|X$ a homeomorphism (of X onto f[X]). Then

$$f[Y \backslash X] \subset Z \backslash f[X].$$

Proof. Let $g : f[X] \to X$ be the homeomorphism inverse to $f|X$, and let $y \in Y$, $f(y) \in f[X]$. Define $h : X \cup \{y\} \to X$ by

$$h(p) = (g \circ f)(p).$$

Then $h|X = id_X$, and hence by Lemma 1.4 (which applies since X is dense in $X \cup \{y\}$) we have $h = id_{X\cup\{y\}}$; it follows that $y = h(y) \in X$. The proof is complete.

1.6. *Lemma.* Let P be a $\cong$-closed class of spaces and let Y and Z be P-reflections of a space X. Then $Y \cong Z$; indeed there is a homeomorphism $h : Y \to Z$ (onto) such that $h(x) = x$ for $x \in X$.

Proof. We define $f : X \to Y$ and $g : X \to Z$ by $f(x) = x$ and $g(x) = x$ for $x \in X$. Since Z and Y are P-reflections of X there are continuous functions $\tilde{f} : Z \to Y$ and $\tilde{g} : Y \to Z$ such that $f \subset \tilde{f}$ and $g \subset \tilde{g}$. Since $\tilde{g} \circ \tilde{f}|X = id_X$ and $\tilde{f} \circ \tilde{g}|X = id_X$ we have $\tilde{g} \circ \tilde{f} = id_Z$ and $\tilde{f} \circ \tilde{g} = id_Y$ by Lemma 1.4. Thus $\tilde{f} = \tilde{g}^{-1}$ and $\tilde{g}$ is a homeomorphism; we set $h = \tilde{g}$.

Remarks. Lemma 1.6 gives a strong sense in which, with P a $\cong$-closed class of spaces and X a space, a P-reflection Y of X (if it exists) is unique. We refer to Y as *the* P-reflection of X and we write $Y = PX$.

When $P = C$ we write $Y = \beta X$ in place of $Y = PX$.

The space βX is the *Stone-Čech* compactification of X.

If P is a class of spaces, and if X and Y are spaces such that X has a P-reflection PX and Y has a P-fication Z, then for every continuous function $f : X \to Y$ we denote by $\tilde{f}$ the continuous function from PX to Z such that $f \subset \tilde{f}$. (The context will always make clear which P-fication of Y we are considering; when none is specified then Y has a P-reflection PY and $\tilde{f}$ denotes the continuous function from PX to PY such that $f \subset \tilde{f}$.) In particular for P a class of spaces, X a space with P-reflection, and continuous $f : X \to Z \in P$, we have $\tilde{f} : PX \to Z$. Again the important special case $P = C$ warrants special notation: for a continuous function $f : X \to Y$ and Z a compactification of Y, the continuous extension of f mapping βX into Z is denoted $\bar{f}$.

We note the following corollary to Lemma 1.6.

1.7. *Corollary.* Let P be a $\cong$-closed class of spaces and let X be a space with a P-reflection. Then $X = PX$ if and only if $X \in P$.

Proof. Since $PX \in P$ we have $X \in P$ if $X = PX$. Conversely if $X \in P$ then it is clear from Lemma 1.6 that X is the (unique) P-reflection of X.

We have seen in Corollary 1.3 that every space has a C-reflection. Using these results we characterize in Theorem 1.11 those classes P such that every space has a P-reflection; 1.8 - 1.10 are preparatory.

We omit the (straightforward) proofs of the following lemmas and corollary.

1.8. *Lemma.* Let Y be a space and $\{A_i : i \in I\}$ a set of subspaces of Y, and set $A = \bigcap_{i\in I} A_i$.

(a) The function $e : A \to \prod_{i\in I} A_i$ defined by $e(x)_i = x$ for $i \in I$

is a homeomorphism from A onto a closed subspace of $\prod_{i \in I} A_i$;

(b) if P is a $\cong$-closed, productive, closed-hereditary class of spaces such that $A_i \in P$ for $i \in I$, then $A \in P$.

1.9. *Lemma.* Let X and Z be spaces and let $f : X \to Z$ be continuous. If $A \subset Z$ then $f^{-1}(A)$ is homeomorphic to a closed subset of $X \times A$.

1.10. *Corollary.* Let P be a $\cong$-closed, finitely productive, closed-hereditary class of spaces and let $f : X \to Z$ be a continuous function. If $X \in P$ and $A \subset Z$ and $A \in P$, then $f^{-1}(A) \in P$.

1.11. *Theorem.* Let P be a class of spaces. The following statements are equivalent.

(a) Every space has a P-reflection;

(b) $C \subset P$, and $P = \bar{P}$.

Proof. (a) ⇒ (b). Every compact space X is dense in the (Hausdorff) space PX, and hence $X = PX \in P$.

We show that P is $\cong$-closed, productive and closed-hereditary.

Let $X' \cong X \in P$ and let $h : X' \to X$ be a homeomorphism (onto). Since $\tilde{h}|X$ is the homeomorphism h we have from Lemma 1.5 that

$$\tilde{h}[PX' \backslash X'] \subset X \backslash X = \emptyset$$

and hence $X' = PX' \in P$. Thus P is $\cong$-closed.

Next let $X_i \in P$ for $i \in I$ and define $X = \prod_{i \in I} X_i$. For $i \in I$ the projection function $\pi_i : X \to X_i$ has a continuous extension $\tilde{\pi}_i : PX \to X_i$, and we define $f : PX \to X$ by

$$f(p)_i = \tilde{\pi}_i(p) \quad \text{for } p \in PX.$$

Since $f|X$ is the homeomorphism id_X and X is dense in

PX we have

$$f[PX \backslash X] \subset X \backslash X = \emptyset$$

by Lemma 1.5, and hence $X = PX$.

Finally let X be any element of P, let A be a closed subset of X, and define $f : A \to X$ by the rule $f(x) = x$ for $x \in A$. We have $\tilde{f} : PA \to X$, and since A is dense in PA we have

$$\tilde{f}[PA] \subset c\ell_X \tilde{f}[A] = c\ell_X A = A,$$

so that

$$\tilde{f}[PA \backslash A] \subset A \backslash A = \emptyset$$

by Lemma 1.5 and hence $A = PA$.

(b) ⇒ (a). For every space X the Stone-Čech compactification βX of X is a C-fication of X. It follows from Theorem 1.2 that X has a $\overline{P}$-reflection. The proof is complete.

We note that if P is a class of spaces such that $C \subset \overline{P}$, then for X a space and $X \subset X' \subset \beta X$ we have $X' \in \overline{P}$ if and only if there are $Z \in \overline{P}$ and a continuous function $f : X \to Z$ such that $X' = \overline{f}^{-1}(Z)$. Indeed if $X' = \overline{f}^{-1}(Z)$ for such f and Z then $X' \in \overline{P}$ by Corollary 1.10; and if $X' \in \overline{P}$ then $X' = \overline{f}^{-1}(Z)$ is as required with $Z = X'$ and $f : X \to X'$ defined by $f(x) = x$. From this remark and Theorem 1.11 we have the following corollary.

1.12. *Corollary*. Let P be a class of spaces such that $C \subset \overline{P}$, let X be a space, and define

$A = \cap\{X' : X \subset X' \subset \beta X$ and $X' \in \overline{P}\}$ and

$B = \cap\{X' :$ there is continuous $f : X \to Z \in \overline{P}$ such that $X' = \overline{f}^{-1}(Z)\}$.

Then X has a $\overline{P}$-reflection $\overline{P}X$, and

$$\overline{P}X = A = B.$$

Proof. Since $A = B$ by the remark above and $A \in \overline{P}$ by Lemma 1.8, it is enough to prove that for every continuous $f : X \to Z \in \overline{P}$ there is a continuous $\tilde{f} : B \to Z$ such that $f \subset \tilde{f}$. Since

$$X \subset B \subset \bar{f}^{-1}(Z) \subset \beta X$$

we may define $\tilde{f} = \bar{f}|B$. The proof is complete.

It is not difficult to give examples proving that in the statement of Corollary 1.12 it is not legitimate to replace "$X' \in \bar{P}$" by "$X' \in P$"; see in this connection the discussion following Corollary 2.4. Nevertheless the space $\bar{P}X$ is determined by continuous functions from X into elements of P, as is shown in Theorem 1.13 and 1.16.

1.13. *Theorem.* Let P be a class of spaces and let X and Y be spaces with X dense in Y. Then X is $\bar{P}$-embedded in Y if and only if X is P-embedded in Y.

Proof. If X is $\bar{P}$-embedded in Y then since $P \subset \bar{P}$ it is clear that X is P-embedded in Y. For the converse suppose that X is P-embedded in Y and let $f : X \to Z \in \bar{P}$ with f continuous. We assume without loss of generality that there is a space $P = \prod_{i \in I} P_i$ such that Z is a closed subspace of P and $P_i \in P$ for $i \in I$. Since $\pi_i \circ f$ is a continuous function from X to P_i there is a continuous function $g_i : Y \to P_i$ such that $\pi_i \circ f \subset g_i$. We define $g : Y \to P$ by

$$g(p)_i = g_i(p).$$

Then g is continuous and $f \subset g$, so to complete the proof it is enough to verify that $g[Y] \subset Z$. Since X is dense in Y we have

$$g[Y] = g[c\ell_Y X] \subset c\ell_P g[X] = c\ell_P f[X] \subset c\ell_P Z = Z,$$

as required.

We note that in Theorem 1.3 the condition that X is dense in Y cannot be omitted. For example if P is the class of arc-wise connected spaces, $X = \{0,1\}$ and $Y = [0,1]$, then X is P-embedded in Y; but X is not $\bar{P}$-embedded since the function $id_X : X \to X \in \bar{P}$ has no continuous extension from Y into X.

1.14. *Corollary.* Let P be a class of spaces and $X \in \bar{P}$. Then there is no space Y such that

X is a proper, dense subset of Y,

Y has a $\bar{P}$-reflection, and

X is P-embedded in Y.

Proof. If such a space Y exists then from Theorem 1.13 it follows that both X and $\bar{P}Y$ are $\bar{P}$-reflections of X. By Lemma 1.6 there is a homeomorphism $h : \bar{P}Y \to X$ such that $h(x) = x$ for $x \in X$, and from Lemma 1.5 we have

$$h[\bar{P}Y \backslash X] \subset X \backslash X = \emptyset$$

and hence $X = \bar{P}Y$. This contradicts the inclusions $X \subsetneq Y \subset PY$.

For classes P such that $C \subset \bar{P}$, Corollary 1.12 describes in two ways how $\bar{P}X$ may be defined in βX "from the outside". In Theorem 1.16 we define $\bar{P}X$ "from the inside", determining explicitly those elements $p \in \beta X$ such that $p \in \bar{P}X$.

1.15. *Lemma*. Let X, Y and Z be spaces with X dense in Y and let $f : X \to Z$ be continuous. In order that there exist a continuous function $g : Y \to Z$ such that $f \subset g$, it is necessary and sufficient that for every $p \in Y$ there is a continuous function $f_p : X \cup \{p\} \to Z$ such that $f \subset f_p$.

Proof. It is enough to verify the sufficiency. Set $g = \bigcup_{p \in Y} f_p$. We prove that g is continuous. If $p \in Y$ and U is a neighborhood in Z of $g(p)$, then there are a neighborhood V of $g(p)$ in Z and a neighborhood W of p in Y such that

$$V \subset c\ell_Z V \subset U \quad \text{and} \quad f[X \cap W] = f_p[X \cap W] \subset V.$$

Now if $y \in W$ we have $y \in c\ell_Y(X \cap W)$ and hence

$$g(y) = f_y(y) \in c\ell_Z f_y[X \cap W] = c\ell_Z f[X \cap W] \subset c\ell_Z V \subset U;$$

it follows that g is continuous at p.

It follows from Lemma 1.15 that if P is a class of spaces and X and Y are spaces with X dense in Y, then X is P-embedded in Y if and only if X is P-embedded in $X \cup \{p\}$ for $p \in Y$.

1.16. *Theorem.* Let P be a class of spaces such that $C \subset \bar{P}$ and let X be a space. Then

$$\bar{P}X = \{p \in \beta X : X \text{ is } P\text{-embedded in } X \cup \{p\}\}.$$

Proof. We set

$$Y = \{p \in \beta X : X \text{ is } P\text{-embedded in } X \cup \{p\}\},$$

and we note from Lemma 1.15 that X is P-embedded in Y. Since $X \subset Y \subset \beta X$ and X is $\bar{P}$-embedded in $\bar{P}X$ we have $\bar{P}X \subset Y$. Further $\bar{P}X$ is P-embedded in Y, since for every continuous function $f : \bar{P}X \to Z \in P$ we have $f|X : X \to Z$ and there is a continuous function $g : Y \to Z$ such that $f|X \subset g$; it is clear that $f \subset g$. Since Y has a $\bar{P}$-reflection by Corollary 1.12, it follows from Corollary 1.14 (with $\bar{P}X$ in place of X) that the inclusion $\bar{P}X \subset Y$ is not a proper inclusion, *i.e.*, that $\bar{P}X = Y$.

1.17. *Corollary.* Let P be a class of spaces such that $C \subset \bar{P}$, and let X be a space.

(a) If $p \in \beta X$, then $p \in \bar{P}X$ if and only if $\bar{f}(p) \in Z$ for every continuous $f : X \to Z \in P$.

(b) $X \in \bar{P}$ if and only if for every $p \in \beta X \setminus X$ there are $Z \in P$ and a continuous $f : X \to Z$ such that $\bar{f}(p) \in \beta Z \setminus Z$.

Proof. For $p \in \beta X$, the condition that there are $Z \in P$ and a continuous $f : X \to Z$ such that $\bar{f}(p) \in \beta Z \setminus Z$ is clearly equivalent to the condition that X is not P-embedded in $X \cup \{p\}$. Thus statement (a) follows from Theorem 1.16. Statement (b) follows from (a) and Corollary 1.6.

§2. Realcompact Spaces and Topologically Complete Spaces

It is easy to give examples of classes of spaces P such that $C \subset \bar{P}$; according to Corollary 1.12 for every such class P every space has a $\bar{P}$-reflection. In these Notes we are concerned in particular with two such classes, which we now define.

Notation. We denote the real line by the symbol R.

Definition. (a) Let $P = \{R\}$. Then $\bar{P}$ is the class of *realcompact* spaces and for every space X the space $\bar{P}$X, denoted υX, is the *Hewitt realcompactification* of X.

(b) Let $P = \{X : X$ is a metrizable space$\}$. Then $\bar{P}$ is the class of *topologically complete* spaces and for every space X the space $\bar{P}$X, denoted γX, is the *Dieudonné topological completion* of X.

Let P be a class of spaces and let X and Y be spaces such that X is P-embedded in Y. It is conventional to say that

X is C*-*embedded* in Y if $P = \{[0,1]\}$;

X is C-*embedded* in Y if $P = \{R\}$; and

X is M-*embedded* in Y if P is the class M of metric spaces.

Thus every space X is C-embedded in υX and M-embedded in γX, and Corollary 1.12 and Theorem 1.16 assume (in part) the following form.

2.1. *Theorem.* Let X be a space. Then

$$\upsilon X = \bigcap\{X' : X \subset X' \subset \beta X \text{ and } X' \text{ is realcompact}\}$$
$$= \{p \in \beta X : X \text{ is C-embedded in } X \cup \{p\}\},$$

and

$$\gamma X = \bigcap\{X' : X \subset X' \subset \beta X \text{ and } X' \text{ is topologically complete}\}$$
$$= \{p \in \beta X : X \text{ is } M\text{-embedded in } X \cup \{p\}\}.$$

For every space X we set

$$C(X) = \{f \in R^X : f \text{ is continuous}\}, \text{ and}$$
$$C^*(X) = \{f \in C(X) : f \text{ is a bounded function}\}.$$

For every $f \in C(X)$ we set

$$Z(f) = f^{-1}(\{0\}) \text{ and } \operatorname{coz} f = X \setminus Z(f);$$

$Z(f)$ and $\operatorname{coz} f$ are called, respectively, the zero-set and the cozero-set of f. We set

$$\mathcal{Z}(X) = \{Z(f) : f \in C(X)\}.$$

The elements of $\mathcal{Z}(X)$ are called *zero-sets* of X; their complements are the *cozero-sets* of X. We note that

$$\mathcal{Z}(X) = \{Z(f) : f \in C^*(X)\}.$$

2.2. *Theorem.* Let X be a space. Then X is realcompact if and only if for every $p \in \beta X \setminus X$ there is a zero-set Z of βX such that $p \in Z \subset \beta X \setminus X$.

Proof. Let X be realcompact and $p \in \beta X \setminus X$. By Corollary 1.17 the space X is not C-embedded in $X \cup \{p\}$, *i.e.*, there is $f \in C(X)$ with no (real-valued) continuous extension to p. It is clear that if U is a neighborhood in βX of p then $f|(X \cap U)$ is unbounded, since otherwise f agrees on $X \cap U$ with a function continuously extendable to p and hence f itself is continuously extendable to p. Now for $x \in X$ define

$$g(x) = 1/[1 + |f(x)|].$$

Then $g \in C^*(X)$, $g(x) > 0$ for $x \in X$, and for every neighborhood U of p in βX the function g assumes values arbitrarily close to 0 on $X \cap U$. Hence $\bar{g}(p) = 0$ and we have

$$p \in \bar{g}^{-1}(\{0\}) \subset \beta X \setminus X.$$

For the converse let $p \in \beta X \setminus X$ and let $f \in C(\beta X)$ be such that $p \in f^{-1}(\{0\}) \subset \beta X \setminus X$. Then f assumes values arbitrarily close to 0 on every set of the form $X \cap U$ with U a neighborhood in βX of p, and hence the function $1/[f|X]$ is unbounded on every such set. This function has no continuous real-valued extension to p and thus X is not C-embedded in $X \cup \{p\}$. It follows that X is realcompact.

Remarks. It is clear from the definitions or from Theorem 2.1 that every realcompact space is topologically complete. Section 6 below (cf. in particular Corollary 6.4 and Remark 6.5) is concerned with the converse. We note that a space X is realcompact if and only if for every $p \in \beta X \setminus X$ there is a continuous function $f : X \cup \{p\} \to R$ such that $f(p) \notin f[X]$, and we observe that (in view of the analogy between realcompact spaces and topologically complete spaces) it is reasonable to conjecture that a space X is topologically complete if and only if for every $p \in \beta X \setminus X$ there are a metric space M and a continuous function $f : X \cup \{p\} \to M$ such that $f(p) \notin f[X]$. It is not difficult to see that this conjecture is equivalent to the statement that every topologically complete space is realcompact. Indeed let X be a space and $p \in \beta X \setminus X$ and suppose that there are a metric space (M,ρ) and a continuous function $f : X \cup \{p\} \to M$ such that $f(p) \notin f[X]$. We set

$$g(q) = \min \{1, \rho(f(q),f(p))\} \text{ for } q \in X \cup \{p\}$$

and we note that g is a continuous function from $X \cup \{p\}$ to $[0,1]$ such that $Z(g) = \{p\}$. Now $\beta X = \beta(X \cup \{p\})$, and we have

$$p \in Z(\bar{g}) \subset \beta X \setminus X.$$

In addition to the two classes of spaces introduced above, we will consider two other classes in some detail in these Notes: the Lindelöf spaces and the paracompact spaces. For the first of these, Theorem 2.3 is analogous to Theorem 2.2.

2.3. *Theorem.* Let X be a space. Then X is a

Lindelöf space if and only if for every compact $K \subset \beta X\backslash X$ there is a zero-set Z of βX such that $K \subset Z \subset \beta X\backslash X$.

Proof. Let X be a Lindelöf space and K a compact subset of $\beta X\backslash X$. For $x \in X$ there is a continuous function $f_x : \beta X \to [0,1]$ such that $f_x(x) = 1$ and $f_x[K] \subset \{0\}$, and we define

$$U_x = f_x^{-1}((1/2,1]) \cap X.$$

Since U_x is open in X and X is Lindelöf there is $\{x_n : n < \omega\} \subset X$ such that $X = \cup_{n<\omega} U_{x_n}$. We set

$$f = \Sigma_{n<\omega} f_{x_n}/2^n$$

and we have $f \in C^*(\beta X)$ and $K \subset Z(f) \subset \beta X\backslash X$.

For the converse let $\{U_i : i \in I\}$ be an open cover of X, for $i \in I$ let V_i be an open subset of βX such that $U_i = X \cap V_i$, and set $K = \cap_{i\in I}(\beta X\backslash V_i)$. Since K is compact and $K \subset \beta X\backslash X$, by assumption there is $f \in C(X)$ such that $K \subset Z(f) \subset \beta X\backslash X$. We note that

$$\text{coz } f \subset \cup_{i\in I} V_i,$$

and that coz f is a σ-compact set (since

$$\text{coz } f = \cup_{1\le n<\omega}\{p \in \beta X : |f(p)| \ge 1/n\}).$$

Thus there is a countable subset J of I such that coz $f \subset \cup_{i\in J} V_i$, and clearly we have

$$X = \cup_{i\in J} U_i.$$

2.4. *Corollary*. Every Lindelöf space is realcompact.

Proof. This follows from Theorems 2.2 and 2.3.

We return briefly to the question given (implicitly) in the paragraph following the proof of Corollary 1.12: If P is a class such that $C \subset \bar{P}$ and X is a space, is it true that for every $p \in \beta X\backslash \bar{P}X$ there is $X' \in P$ such that $X \subset X' \subset \beta X\backslash\{p\}$? It follows from Corollary 2.4 that if $P = \{R\}$ and X is a Lindelöf space then $X = \bar{P}X$; if in addition X is not compact and not metrizable, then $\beta X\backslash \bar{P}X \neq \emptyset$ and for $p \in \beta X\backslash \bar{P}X$ there is no $X' \in P$ such that $X \subset X' \subset \beta X\backslash\{p\}$.

The lemma and theorem which follow are preparatory for Section 6, where the precise relationship between realcompact and topologically complete spaces is examined.

Definition. A cardinal α is *Ulam-measurable* if there is a function $\mu : \mathcal{P}(\alpha) \to \{0,1\}$ such that

$\mu(\emptyset) = 0,$

$\mu(\alpha) = 1,$

$\mu(\{\xi\}) = 0$ for $\xi < \alpha$, and

if $\mathcal{A}$ is a countable family of pairwise disjoint subsets of α, then

$$\mu(\cup \mathcal{A}) = \Sigma\{\mu(A) : A \in \mathcal{A}\}.$$

Such a function μ is called an *Ulam measure* on α. More generally, if S is a set and μ a function from $\mathcal{P}(S)$ to $\{0,1\}$, then μ is said to be an Ulam measure on S if it satisfies (the obvious modifications of) the above conditions. Clearly there is an Ulam measure on S if and only if $|S|$ is an Ulam-measurable cardinal.

We note that if μ is an Ulam measure on α and $A \subset B \subset \alpha$, then

$$\mu(A) + \mu(B\backslash A) = \mu(B)$$

and hence $\mu(A) \leq \mu(B)$.

It is clear that every Ulam-measurable cardinal is uncountable. The following lemma proves that the smallest Ulam-measurable cardinal (if it exists) must be very large.

2.5. *Lemma.* (a) If α is a cardinal that is not Ulam-measurable and $\beta < \alpha$, then β is not Ulam-measurable;

(b) if $|I|$ is not Ulam-measurable and $\{\alpha_i : i \in I\}$ is a set of non-Ulam-measurable cardinals, then $\Sigma_{i\in I}\, \alpha_i$ is not Ulam-measurable; and

(c) if α is a cardinal that is not Ulam-measurable,

then 2^α is not Ulam-measurable.

Proof. (a) If μ is an Ulam measure on β and ν is defined on $P(\alpha)$ by the rule $\nu(A) = \mu(A \cap \beta)$, then ν is an Ulam measure on α.

(b) Set $\alpha = \Sigma_{i \in I} \alpha_i$ and let $\{A_i : i \in I\}$ be a family of pairwise disjoint subsets of α such that $|A_i| = \alpha_i$ and $\cup_{i \in I} A_i = \alpha$. If μ is an Ulam measure on α then $\mu(A_i) = 0$ for $i \in I$, since otherwise $\mu | P(A_i)$ is an Ulam measure on A_i. We define $\nu : P(I) \to \{0,1\}$ by

$$\nu(J) = \mu(\cup\{A_i : i \in J\})$$

and we note that ν is an Ulam measure on I; hence $|I|$ is Ulam-measurable.

(c) Suppose instead that

$$\mu : P(P(\alpha)) \to \{0,1\}$$

is an Ulam measure on $P(\alpha)$, for $\xi < \alpha$ define

$$S_\xi = \{A \in P(\alpha) : \xi \in A\} \text{ and } S'_\xi = \{A \in P(\alpha) : \xi \notin A\},$$

and set $S = \{\xi < \alpha : \mu(S_\xi) = 1\}$. We note that $\mu(S'_\xi) = 1$ for $\xi \in \alpha \backslash S$, and we note also from (b), using the fact that α is not Ulam-measurable, that the union of any α subsets of $P(\alpha)$ with μ-measure 0 is again of μ-measure 0. Thus

$$\mu[(\cap_{\xi \in S} S_\xi) \cap (\cap_{\xi \in \alpha \backslash S} S'_\xi)] = 1,$$

i.e., $\mu(\{S\}) = 1$ and μ is not an Ulam measure on $P(\alpha)$.

2.6. *Theorem.* For every cardinal number α the following statements are equivalent.

(a) The discrete space α is realcompact;

(b) α is not Ulam-measurable.

Proof. (a) $\Rightarrow$ (b). If μ is an Ulam measure on α then $\{A \subset \alpha : \mu(A) = 1\}$ has the countable intersection property, hence *a fortiori* the finite intersection property, and there is $p \in \cap\{cl_{\beta(\alpha)} A : \mu(A) = 1\}$. We note that if $A \subset \alpha$ and $p \in cl_{\beta(\alpha)} A$ then $\mu(A) = 1$; for otherwise

$\mu(\alpha\backslash A) = 1$ and the function $\chi_A : \alpha \to \{0,1\}$ defined by the rule

$$\chi_A(\xi) = 1 \quad \text{if} \quad \xi \in A$$
$$= 0 \quad \text{if} \quad \xi \in \alpha\backslash A$$

has no continuous extension to p.

Since $\mu(\{\xi\}) = 0$ for $\xi < \alpha$ we have $p \in \beta(\alpha)\backslash\alpha$ and since α is realcompact there is by Theorem 3.2 a continuous function $f : \beta(\alpha) \to [0,1]$ such that $p \in Z(f) \subset \beta(\alpha)\backslash\alpha$. For $n < \omega$ we define

$$U_n = \{q \in \beta(\alpha) : f(q) < 1/n\}$$

and we note that U_n is a neighborhood of p and hence $p \in c\ell_{\beta(\alpha)}(U_n \cap \alpha)$. Then $\mu(U_n \cap \alpha) = 1$ for $n < \omega$, and hence

$$\mu(\cap_{n<\omega}(U_n \cap \alpha)) = 1;$$

but $\cap_{n<\omega}(U_n \cap \alpha) = Z(f) \cap \alpha = \emptyset$.

(b) $\Rightarrow$ (a). Let $p \in \beta(\alpha)\backslash\alpha$ and define $\mu : P(\alpha) \to \{0,1\}$ by

$$\mu(A) = 1 \quad \text{if} \quad p \in c\ell_{\beta(\alpha)} A$$
$$= 0 \quad \text{if} \quad p \notin c\ell_{\beta(\alpha)} A.$$

Clearly, $\mu(\alpha) = 1$, $\mu(\{\xi\}) = 0$ for $\xi < \alpha$, and μ is finitely additive; thus (since μ is not an Ulam measure on α) there is $\{A_n : n < \omega\} \subset P(\alpha)$ such that $p \notin c\ell_{\beta(\alpha)} A_n$ for $n < \omega$ but $p \in c\ell_{\beta(\alpha)}(\cup_{n<\omega} A_n)$. We set $A = \cup_{n<\omega} A_n$ and we note that $p \notin c\ell_{\beta(\alpha)}(\alpha\backslash A)$ (since otherwise the function $\chi_A : \alpha \to \{0,1\}$ has no continuous extension to p). There is a continuous function $f : \beta(\alpha) \to [0,1]$ such that $f(p) = 0$ and $f(\xi) = 1$ for $\xi \in \alpha\backslash A$, and for $n < \omega$ there is a continuous function $g_n : \beta(\alpha) \to [0,1]$ such that $g_n(p) = 0$ and $g_n(\xi) = 1$ for $\xi \in A_n$. We define

$$h = f + \Sigma_{n<\omega}\, g_n/2^n$$

and we have $h \in C(\beta(\alpha))$ and

$$p \in Z(h) \subset \beta(\alpha)\backslash\alpha.$$

It follows from Theorem 2.2 that the space α is realcompact.

We conclude this Section with a characterization of topologically complete spaces in terms of products of complete metric spaces (analogous to the characterization of realcompact spaces in terms of powers of R).

2.7. *Theorem.* For every space X, the following statements are equivalent.

(a) X is topologically complete;

(b) X is homeomorphic with a closed subset of product of complete metric spaces.

Proof. It is enough to verify that (a) ⇒ (b), and for this it is enough to prove that every metric space M is homeomorphic with a closed subset of a product of complete metric spaces. Let $\bar{M}$ denote the metrizable completion of M and define

$$M_p = \bar{M}\backslash\{p\} \quad \text{for} \quad p \in \bar{M}\backslash M.$$

As is well-known (and as we prove in Lemma 3.4 below) every space M_p has a compatible complete metric. Further the function $e : M \to \prod_{p\in\bar{M}\backslash M} M_p$ defined by

$$e(x)_p = x \quad \text{for} \quad x \in M$$

is a homeomorphism from M onto a closed subset of $\prod_{p\in\bar{M}\backslash M} M_p$ (by Lemma 1.8(a)). The proof is complete.

§3. METRIC SPACES

In this Section we give three results about metric spaces. The first (Theorem 3.1) is A. H. Stone's classical theorem that every metric space is paracompact; the second (Theorem 3.2) allows for the definition of sufficiently many continuous pseudometrics on certain spaces; and the third (Theorem 3.6) is E. Čech's classical characterization of those metric spaces M such that M is a G_δ in βM. The three results are unrelated (except that each concerns metric spaces), but they are indispensable for later use.

In these Notes we shall not distinguish carefully between a metric space (M,ρ) and a metrizable space M. When a symbol ρ or σ appears in a proof concerning a metric space or a metrizable space M, it is understood to denote a compatible metric function for M.

For every metric space M, $p \in M$ and $\varepsilon > 0$, we set

$$S(p,\varepsilon) = \{x \in M : \rho(x,p) < \varepsilon\}.$$

Definitions. Let X be a space and $\mathcal{A}$ a family of subsets of X.

(a) $\mathcal{A}$ is *locally finite (in* X*)* if for every $p \in X$ there is a neighborhood U of p such that

$$|\{A \in \mathcal{A} : U \cap A \neq \emptyset\}| < \omega;$$

and $\mathcal{A}$ is *discrete (in* X*)* if for every $p \in X$ there is a neighborhood U of p such that

$$|\{A \in \mathcal{A} : U \cap A \neq \emptyset\}| \leq 1;$$

(b) a family $\mathcal{B}$ of subsets of X is a *refinement* of $\mathcal{A}$ if $\cup\mathcal{B} = \cup\mathcal{A}$ and for every $B \in \mathcal{B}$ there is $A \in \mathcal{A}$ such that $B \subset A$;

(c) X is *paracompact* if every open cover of X has a locally finite open refinement.

3.1. *Theorem.* Let (M,ρ) be a metric space. Then

(a) M is paracompact; and

(b) every open cover of M has a refinement that is both locally finite and σ-discrete (*i.e.*, a countable union of discrete families).

Proof. It is enough to prove (b). Let $\mathcal{U} = \{U_\xi : \xi < \alpha\}$ be an open cover of M, for $\xi < \alpha$ let $V_{\xi,0}$ denote the union of all sets $S(x,1)$ such that

ξ is the smallest ordinal such that $x \in U_\xi$, and
$S(x,3) \subset U_\xi$;

and recursively, if $V_{\xi,k}$ has been defined for all $\xi < \alpha$ and $k < n$, let $V_{\xi,n}$ denote the union of all sets $S(x,1/2^n)$ such that

ξ is the smallest ordinal such that $x \in U_\xi$,
$S(x,3/2^n) \subset U_\xi$, and
$x \notin \cup\{V_{\eta,k} : \eta < \alpha,\ k < n\}$.

We define $\mathcal{V}_n = \{V_{\xi,n} : \xi < \alpha\}$ and $\mathcal{V} = \cup_{n<\omega} \mathcal{V}_n$. We note that if $x \in M$ and ξ is the smallest ordinal such that $x \in U_\xi$, then there is $n < \omega$ such that $S(x,3/2^n) \subset U_\xi$, so that either $x \in V_{\xi,n}$ or there are $k < n$ and $\eta < \alpha$ such that $x \in V_{\eta,k}$. Thus $\cup\{V_{\xi,n} : \xi < \alpha,\ n < \omega\} = M$, and since $V_{\xi,n} \subset U_\xi$ the family $\mathcal{V}$ is a refinement of $\mathcal{U}$.

We claim next that $\mathcal{V}_n$ is discrete for $n < \omega$. Indeed let $p \in M$ and suppose that there are ξ and η such that $\xi < \eta < \alpha$ and the neighborhood $S(p,1/2^{n+1})$ of p intersects both $V_{\xi,n}$ and $V_{\eta,n}$. If

$u \in S(p,1/2^{n+1}) \cap V_{\xi,n}$ and $v \in S(p,1/2^{n+1}) \cap V_{\eta,n}$

then there are $x, y \in M$ such that

$u \in S(x,1/2^n) \subset S(x,3/2^n) \subset U_\xi$,
$v \in S(y,1/2^n) \subset S(y,3/2^n) \subset U_\eta$, and
$y \notin U_\xi$;

it follows that

$$\rho(x,y) \le \rho(x,u) + \rho(u,p) + \rho(p,v) + \rho(v,y)$$
$$< 1/2^n + 1/2^{n+1} + 1/2^{n+1} + 1/2^n = 3/2^n$$

and hence $y \in U_\xi$, a contradiction.

We prove finally that the cover $\mathcal{V}$ is locally finite. If $p \in M$ and $\xi < \alpha$, $m < \omega$ are chosen so that $p \in V_{\xi,m}$, then since $V_{\xi,m}$ is open there is $n < \omega$ such that $n \ge m$ and

$$p \in S(p,1/2^n) \subset V_{\xi,m};$$

from the paragraph above it follows that $S(p,1/2^{n+1})$ intersects at most one element of $\mathcal{V}_k$ if $k \le n$. And if $k > n$ then $S(p,1/2^{n+1}) \cap V_{\eta,k} = \emptyset$ for $\eta < \alpha$, since if $u \in S(p,1/2^{n+1}) \cap V_{\eta,k}$ there is

$$x \in X \backslash \cup\{V_{\zeta,i} : \zeta < \alpha,\ i < k\}$$

such that $u \in S(x,1/2^k)$ and since in particular $x \notin V_{\xi,m}$ we have

$$1/2^n \le \rho(x,p) \le \rho(x,u) + \rho(u,p)$$
$$< 1/2^k + 1/2^{n+1} \le 1/2^{n+1} + 1/2^{n+1} = 1/2^n,$$

a contradiction. Thus $S(p,1/2^{n+1})$ is a neighborhood of p that intersects at most $n + 1$ elements of $\mathcal{V}$. The proof is complete.

3.2. *Theorem.* Let X be a space and let $\mathcal{U} = \cup_{n<\omega} \mathcal{U}_n$ be a cozero cover of X such that for every $n < \omega$ the family $\mathcal{U}_n$ is locally finite in X. For $U \in \mathcal{U}$ let f_U be a continuous function from X into $[0,1]$ such that $U = \operatorname{coz} f_U$, for $n < \omega$ define $d_n : X \times X \to R$ by the rule

$$d_n(x,y) = \min \{1, \Sigma\{|f_U(x) - f_U(y)| : U \in \mathcal{U}_n\}\},$$

and define

$$d = \Sigma_{n<\omega}\, d_n/2^n.$$

For $x \in X$ set $\bar{x} = \{y \in X : d(x,y) = 0\}$ and define

$$M = \{\bar{x} : x \in X\}, \text{ and } \rho(\bar{x},\bar{y}) = d(x,y) \text{ for } \bar{x}, \bar{y} \in M.$$

Then

(a) each of the functions d_n, and the function d, are well-defined functions from $X \times X$ to R;

(b) ρ is a (well-defined) metric for M;

(c) the function $f : X \to M$ defined by

$$f(x) = \bar{x} \text{ for } x \in M$$

is a continuous function from X onto the metric space (M,ρ); and

(d) if $p \in \beta X\backslash X$ and there is a cover $\mathcal{V}(p)$ of X such that $\mathcal{V}(p) \subset \mathcal{U}$ and

$$p \notin \cup\{c\ell_{\beta X}\, U : U \in \mathcal{V}(p)\},$$

and if BM is a compactification of M, then $\bar{f}(p) \in BM\backslash M$ (where $\bar{f}$ denotes the continuous function from βX onto BM such that $f \subset \bar{f}$).

Proof. (a) For $n < \omega$ the family $\mathcal{U}_n$ is locally finite, so for every $x \in X$ there are only finitely many $U \in \mathcal{U}$ such that $f_U(x) \neq 0$. Thus $d_n : X \times X \to R$ is well-defined, and hence d is well-defined.

(b) For $x, x' \in X$ we have $\bar{x} = \overline{x'}$ if and only if $f_U(x) = f_U(x')$ for every $U \in \mathcal{U}$. It follows that ρ is a well-defined function from $M \times M$ to R, and that $\rho(\bar{x},\bar{y}) = 0$ if and only if $\bar{x} = \bar{y}$. The remaining verifications required to prove (b) are routine and are omitted.

(c) Let $x \in X$ and $\varepsilon > 0$, and let $k < \omega$ be such that $1/2^{k-1} < \varepsilon/2$. Since $\mathcal{U}_n$ is locally finite for $n < k$, there are a neighborhood V of x and $m < \omega$ such that

$$|\{U : U \in \cup_{n<k} \mathcal{U}_n,\ V \cap U \neq \emptyset\}| = m,$$

and there is a neighborhood W of x such that if $y \in W$ and $U \in \cup_{n<k} \mathcal{U}_n$ and $V \cap U \neq \emptyset$ then

$$|f_U(x) - f_U(y)| < \varepsilon/(2\cdot k\cdot m).$$

Then for $y \in V \cap W$ and $n < k$ we have

$$d_n(x,y) \leq \Sigma\{|f_U(x) - f_U(y)| : U \in \mathcal{U}_n\}$$
$$\leq m\cdot\varepsilon/(2\cdot k\cdot m) = \varepsilon/(2\cdot k),$$

and hence for $y \in V \cap W$ we have

$$\rho(f(x),f(y)) = d(x,y) \leq \Sigma_{n<k}\, \varepsilon/(2\cdot k\cdot 2^n) + \Sigma_{n\geq k}\, 1/2^n$$
$$< \varepsilon/2 + \varepsilon/2 = \varepsilon.$$

(d) If (d) fails there are $p \in \beta X\backslash X$, $x \in X$, $n < \omega$ and $U \in \mathcal{U}_n \cap \mathcal{V}(p)$ such that

$$f(x) = \bar{f}(p), \quad x \in U, \quad \text{and} \quad p \notin c\ell_{\beta X} U.$$

We define $\eta = f_U(x)$ and we choose an open neighborhood W of $\bar{f}(p)$ in BM such that

$$W \cap M = \{z \in M : \rho(\bar{f}(p),z) < \eta/2^n\}.$$

Since $p \notin c\ell_{\beta X} U = c\ell_{\beta X} (\text{coz } f_U)$ we have $p \in c\ell_{\beta X} Z(f_U)$ and hence there is $y \in \bar{f}^{-1}(W) \cap Z(f_U)$. Then $\bar{f}(y) \in W$ and hence

$$\rho(\bar{f}(p),\bar{f}(y)) < \eta/2^n;$$

but

$$\rho(\bar{f}(p),\bar{f}(y)) = \rho(f(x),f(y))$$
$$\geq |f_U(x) - f_U(y)|/2^n = f_U(x)/2^n = \eta/2^n,$$

a contradiction. The proof is complete.

Definition. Let Y be a space and $G \subset Y$. Then G is a G_δ subset of Y if there is a sequence $\{G_n : n < \omega\}$ of open subsets of Y such that $G = \bigcap_{n<\omega} G_n$.

If (X,ρ) is a metric space and $A \subset X$, we set

$$\rho\text{-diam } A = \sup \{\rho(x,y) : x, y \in A\} \quad \text{if } A \neq \emptyset$$
$$= 0 \quad \text{if } A = \emptyset.$$

3.3. *Lemma.* Let Y be a space and X a dense subspace of Y, and let f be a continuous function from X into a complete metric space (M,ρ). Then there are a G_δ subset G of Y and a continuous function $\tilde{f} : G \to M$ such that

$$X \subset G \quad \text{and} \quad \tilde{f}|X = f.$$

Proof. For $1 \leq n < \omega$ we set

$$G_n = \{y \in Y : y \text{ has a neighborhood } U \text{ such that } \rho\text{-diam } f[U \cap X] < 1/n\}$$

and we set

$$G = \bigcap_{1\leq n<\omega} G_n.$$

Clearly G_n is open in Y, and $X \subset G$. If $y \in G$ then for $1 \leq n < \omega$ there is a neighborhood U_n of y such that $\rho\text{-diam } f[U_n \cap X] < 1/n$, and since (M,ρ) is a complete metric space there is $p \in M$ such that

$$\bigcap_{1\leq n<\omega} c\ell_M f[U_n \cap X] = \{p\};$$

we set $f_y = f \cup \{<y,p>\}$ and we claim that f_y is continuous. Indeed if $\varepsilon > 0$ there is n such that $1 \le n < \omega$ and $1/n < \varepsilon$, and since

$$f_y(y) \in c\ell_M\, f[U_n \cap X]$$

we have

$$\rho(f_y(x), f_y(y)) = \rho(f(x), f_y(y)) \le 1/n < \varepsilon \quad \text{for } x \in U_n \cap X.$$

We define $\tilde{f} = \cup_{y \in G} f_y$; it follows from Lemma 1.15 (with Y and Z replaced by G and M respectively) that $\tilde{f}$ is as required.

Definition. A metric space (M,ρ) is *completely metrizable* if there is a compatible metric σ for M such that (M,σ) is a complete metric space.

3.4. *Lemma.* An open subset of a complete metric space is completely metrizable.

Proof. Let (M,ρ) be a complete metric space and let U be open in M. For $x \in U$ we define

$$f(x) = 1/\rho(x, M\backslash U)$$

and for $x, y \in U$ we define

$$\sigma(x,y) = \rho(x,y) + |f(x) - f(y)|.$$

It is clear that

$$\sigma(x,y) = \sigma(y,x),$$

$$\sigma(x,y) = 0 \text{ if and only if } x = y, \text{ and}$$

$$\sigma(x,z) \le \sigma(x,y) + \sigma(y,z)$$

for $x, y, z \in U$,

and hence σ is a metric for the set U. Further, if $\{x_n : n < \omega\} \subset U$ and $x \in U$, then $\sigma(x_n,x) \to 0$ if and only if $\rho(x_n,x) \to 0$; thus σ is compatible with ρ on U.

It remains to prove that (U,σ) is a complete metric space. Let $\{x_n : n < \omega\}$ be a σ-Cauchy sequence of elements of U. Then $\{x_n : n < \omega\}$ is a ρ-Cauchy sequence; and $\{f(x_n) : n < \omega\}$ is Cauchy in the usual metric of R. Hence (since (M,ρ) is a complete metric space) there is $x \in M$ such that $\rho(x_n,x) \to 0$; and $\{f(x_n) : n < \omega\}$ is bounded, so that $x \in U$. Since σ is compatible with ρ on U, it follows that $\sigma(x_n,x) \to 0$. The proof is complete.

3.5. *Lemma.* For every space X, the following statements are equivalent.

(a) X is a G_δ in βX;

(b) X is a G_δ in every compactification of X;

(c) X is a G_δ in some compactification of X.

Proof. (a) ⇒ (b). Let BX be a compactification of X. Since βX is the C-reflection of X there is a continuous function $f : \beta X \to BX$ such that $f(x) = x$ for every $x \in X$, and from Lemma 1.5 we have

$$f[\beta X\backslash X] \subset BX\backslash f[X] = BX\backslash X.$$

Since $f[\beta X] = BX$ it follows that $f[\beta X\backslash X] = BX\backslash X$. Since $\beta X\backslash X$ is σ-compact the space $BX\backslash X$ is σ-compact, and hence X is a G_δ in BX.

(b) ⇒ (c). This is obvious.

(c) ⇒ (a). Let BX be a compactification of X in which X is a G_δ. As in the proof above that (a) ⇒ (b) there is a continuous function $f : \beta X \to BX$ such that

$$f[X] = X \text{ and } f[\beta X\backslash X] = BX\backslash X.$$

There is a countable family $\{U_n : n < \omega\}$ of open subsets of BX such that $X = \bigcap_{n<\omega} U_n$, and hence in βX we have

$$X = f^{-1}(X) = f^{-1}(\textstyle\bigcap_{n<\omega} U_n) = \textstyle\bigcap_{n<\omega} f^{-1}(U_n),$$

as required.

3.6. *Theorem.* For every metrizable space M, the following statements are equivalent.

(a) M is completely metrizable;

(b) M is a G_δ in βM;

(c) M is a G_δ in some complete metric space.

Proof. (a) ⇒ (b). By Lemma 3.3 (applied to the function $f = id_M : M \to M$) there are a G_δ subset G of

βM and a continuous function $\tilde{f} : G \to M$ such that $\tilde{f}(x) = x$ for $x \in M$. From Lemma 1.5 we have

$$\tilde{f}[G \backslash M] \subset M \backslash M = \emptyset,$$

and hence $M = G$.

(b) ⇒ (c). Let $\bar{M}$ denote the usual metrizable completion of M. Then $\beta\bar{M}$ is a compactification of M, and $M \subset \bar{M} \subset \beta\bar{M}$. It follows from Lemma 3.5 that M is a G_δ in $\beta\bar{M}$; it is then clear that M is a G_δ in $\bar{M}$.

(c) ⇒ (a). Let X be a complete metric space and $\{G_n : n < \omega\}$ a sequence of open subsets of X such that $M = \bigcap_{n<\omega} G_n$. By Lemma 3.4 the space G_n is completely metrizable for $n < \omega$, and hence $\prod_{n<\omega} G_n$ is completely metrizable. Since M is homeomorphic with a closed subspace of $\prod_{n<\omega} G_n$ (by Lemma 1.8(a)), the space M is completely metrizable.

§4. X AS A SUBSET OF βX; LOCALLY FINITE COVERS

We have seen in Theorem 2.3 that X is a Lindelöf space if and only if for every compact $K \subset \beta X \backslash X$ there is $Z \in \mathcal{Z}(\beta X)$ such that $K \subset Z \subset \beta X \backslash X$. In Theorem 4.4 below we sharpen this result and obtain characterizations of realcompact, of paracompact, and of topologically complete spaces in the same spirit. These characterizations are given in terms of the existence of locally finite cozero covers of X.

We note that it follows from our definition of space that every open set is a union of cozero sets, *i.e.*, every open cover has a cozero refinement.

It is easy to show that if X is a space and $\mathcal{A}$ is a locally finite family of subsets of X, then

$$c\ell_X(\cup\{A : A \in \mathcal{A}\}) = \cup \{c\ell_X A : A \in \mathcal{A}\}.$$

4.1. *Lemma.* For every space X, every countable cozero cover of X has a countable, locally finite, cozero refinement.

Proof. Let $\{f_n : n < \omega\}$ be a sequence of elements of $C^*(X)$ such that

$$X = \cup_{n<\omega} \operatorname{coz} f_n$$

and for $n < \omega$ define

$$U_{n,k} = \{x \in X : |f_n(x)| > 1/k\} \text{ and}$$
$$Z_{n,k} = \{x \in X : |f_n(x)| \geq 1/k\}$$

for $k < \omega$. It is clear that $U_{n,k}$ and $Z_{n,k}$ are cozero-sets and zero-sets respectively of X; we define

$$V_{n,k} = U_{n,k} \backslash \cup \{Z_{i,j} : i + j \leq n + k - 2\}$$

and we note that $V_{n,k}$ is a cozero-set of X. We claim that $\{V_{n,k} : n < \omega, k < \omega\}$ is a cover for X. Indeed let

$x \in X$ and among all pairs $\{i,j\}$ with $i < \omega$, $j < \omega$ such that $x \in U_{i,j}$ let $\{n,k\}$ have minimal sum; then $x \in U_{n,k}$ and in fact $x \in V_{n,k}$: otherwise there is $\{i,j\}$ such that $x \in Z_{i,j}$ with $i + j \leq n + k - 2$, and then $x \in U_{i,j+1}$ and $i + (j + 1) < n + k$.

To complete the proof it is enough to show that if $\langle i,j \rangle \in \omega \times \omega$ then

$$|\{\langle n,k \rangle : V_{i,j} \cap V_{n,k} \neq \emptyset\}| < \omega.$$

If $V_{i,j} \cap V_{n,k} \neq \emptyset$ then $Z_{i,j} \cap V_{n,k} \neq \emptyset$, and hence $n + k < i + j + 2$.

4.2. *Lemma.* (a) Every Lindelöf space is paracompact;

(b) every paracompact space is normal.

Proof. (a) follows from Lemma 4.1.

(b) Let X be paracompact and let A and B be disjoint closed subsets of X. For every $x \in A$ there is an open neighborhood U_x of x such that $cl_X U_x \cap B = \emptyset$; we define

$$\mathcal{U} = \{U_x : x \in A\} \cup \{X \backslash A\}.$$

Then $\mathcal{U}$ is an open cover of X and there is a locally finite open refinement $\mathcal{V}$ of $\mathcal{U}$. We set

$$U = \cup \{V \in \mathcal{V} : V \cap A \neq \emptyset\}$$

and we see that U and $X \backslash cl_X U$ are disjoint neighborhoods in X of A and B respectively.

A family $\mathcal{A}$ of subsets of a space X is said to be *point-finite* (in X) if

$$|\{A \in \mathcal{A} : p \in A\}| < \omega \quad \text{for } p \in X.$$

4.3. *Lemma.* For every space X the following are equivalent.

(a) X is paracompact;

(b) every open cover of X has a locally finite cozero refinement.

Proof. It is sufficient to prove that (a) $\Rightarrow$ (b). We begin with the following observation.

(*) For every normal space Y and every point-finite open cover $\{U_i : i \in I\}$ of Y there is an open cover $\{V_i : i \in I\}$ of Y such that

$$V_i \subset c\ell_Y V_i \subset U_i \quad \text{for all } i \in I.$$

To prove (*) we denote by M the set of all pairs $\langle J,\phi\rangle$ with $J \subset I$ and ϕ a function from J to the family of open subsets of X such that

$\{\phi(i) : i \in J\} \cup \{U_i : i \in I\backslash J\}$ is a cover for X, and

$\phi(i) \subset c\ell_Y \phi(i) \subset U_i$ for $i \in J$.

A relation $\leq$ is defined on M by the rule

$$\langle J,\phi\rangle \leq \langle K,\psi\rangle \text{ if } J \subset K \text{ and } \psi|J = \phi.$$

It is clear that $\langle M,\leq\rangle$ is a partially ordered set, and it is easy to show (using the fact that the cover $\{U_i : i \in I\}$ is point-finite) that the union of a chain in $\langle M,\leq\rangle$ is an element of M and an upper bound for the chain. Since $\langle\emptyset,\emptyset\rangle \in M$ we have $M \neq \emptyset$ and hence by Zorn's Lemma there is a maximal element $\langle J,\phi\rangle$ of $\langle M,\leq\rangle$. We claim that $J = I$. If there is $\bar{i} \in I\backslash J$ we define

$$A = Y\backslash[\cup\{\phi(i) : i \in J\} \cup \cup\{U_i : i \notin J,\ i \neq \bar{i}\}]$$

and we note that since $A \subset U_{\bar{i}}$ and A is closed in Y there is (by the normality of Y) an open set V such that

$$A \subset V \subset c\ell_Y V \subset U_{\bar{i}};$$

then $\langle J \cup \{\bar{i}\}, \phi \cup \{\langle\bar{i},V\rangle\}\rangle$ is an element of M greater than $\langle J,\phi\rangle$, a contradiction. The proof of (*) is complete.

To complete the proof that (a) $\Rightarrow$ (b) let $\mathcal{A}$ be an open cover of X, let $\mathcal{U} = \{U_i : i \in I\}$ be a locally finite open refinement of $\mathcal{A}$, and (using Lemma 4.2(b)) let $\{V_i : i \in I\}$ be as given by (*). Since X is normal there is a family $\{f_i : i \in I\}$ of continuous functions from X into $[0,1]$ such that

$$f_i[V_i] \subset \{1\} \text{ and } f_i[X\backslash U_i] \subset \{0\} \text{ for } i \in I,$$

and it is clear that the family $\{\text{coz } f_i : i \in I\}$ is a locally finite, cozero refinement of $\mathcal{A}$.

4.4. *Theorem.* Let X be a space. Then

(a) X is a Lindelöf space if and only if for every compact $K \subset \beta X \backslash X$ there is a countable, locally finite cozero cover $\mathcal{U}$ of X such that $U \in \mathcal{U}$ implies $K \cap c\ell_{\beta X} U = \emptyset$;

(b) X is realcompact if and only if for every $p \in \beta X \backslash X$ there is a countable, locally finite cozero cover $\mathcal{U}$ of X such that $U \in \mathcal{U}$ implies $p \notin c\ell_{\beta X} U$;

(c) X is paracompact if and only if for every compact $K \subset \beta X \backslash X$ there is a locally finite cozero cover $\mathcal{U}$ of X such that $U \in \mathcal{U}$ implies $K \cap c\ell_{\beta X} U = \emptyset$; and

(d) X is topologically complete if and only if for every $p \in \beta X \backslash X$ there is a locally finite cozero cover $\mathcal{U}$ of X such that $U \in \mathcal{U}$ implies $p \notin c\ell_{\beta X} U$.

Proof. (a) If X is a Lindelöf space and K is a compact subset of $\beta X \backslash X$ then by Theorem 2.3 there is $f \in C(\beta X)$ such that $0 < f(x) < 1$ for all $x \in X$ and $K \subset Z(f) \subset \beta X \backslash X$. For $n < \omega$ we define

$$U_n = \{x \in X : 1/(n + 3) < f(x) < 1/(n + 1)\},$$

and we set $\mathcal{U} = \{U_n : n < \omega\}$. It is clear that $\mathcal{U}$ is as required.

For the converse let

K be a compact subset of $\beta X \backslash X$,

$\{f_n : n < \omega\} \subset C(X)$ be such that $0 \leq f_n(x) \leq 1$ for $n < \omega$ and $x \in X$,

$\{\text{coz } f_n : n < \omega\}$ a locally finite cover of X, and

$K \cap (\cup\{c\ell_{\beta X} \text{ coz } f_n : n < \omega\}) = \emptyset$.

Then $K \subset c\ell_{\beta X} Z(f_n)$ for $n < \omega$ and we have $\overline{f_n}(p) = 0$ for $n < \omega$ and $p \in K$. We define

$$f = \Sigma_{n<\omega} \overline{f_n}/2^n$$

and we have $f \in C(\beta X)$ and $K \subset Z(f) \subset \beta X \backslash X$. It follows from Theorem 2.3 that X is a Lindelöf space.

(b) This proof is analogous to the proof of (a);

instead of Theorem 2.3 we use Theorem 2.2, which states that X is realcompact if and only if for every $p \in \beta X \backslash X$ there is $f \in C(\beta X)$ such that $p \in Z(f) \subset \beta X \backslash X$.

(c) Let X be paracompact and K a compact subset of $\beta X \backslash X$, and for $x \in X$ let V_x be an open subset of X such that

$$K \cap c\ell_{\beta X} V_x = \emptyset.$$

By Lemma 4.3 the open cover $\{V_x : x \in X\}$ has a locally finite cozero refinement $\mathcal{U}$; it is clear that $\mathcal{U}$ is as required.

For the converse suppose that for every compact $K \subset \beta X \backslash X$ there is a locally finite open cover $\mathcal{U}$ of X such that $U \in \mathcal{U}$ implies $K \cap c\ell_{\beta X} U = \emptyset$, and let $\mathcal{V}$ be an open cover of X. For $V \in \mathcal{V}$ there is an open subset $\tilde{V}$ of βX such that $\tilde{V} \cap X = V$. We define

$$K = \cap\{\beta X \backslash \tilde{V} : V \in \mathcal{V}\}$$

and we choose a locally finite cozero cover $\mathcal{U}$ of X such that $U \in \mathcal{U}$ implies $K \cap c\ell_{\beta X} U = \emptyset$. Then $c\ell_{\beta X} U \subset \cup\{\tilde{V} : V \in \mathcal{V}\}$, so since $c\ell_{\beta X} U$ is compact there is a finite family $\mathcal{V}(U) \subset \mathcal{V}$ such that

$$c\ell_{\beta X} U \subset \cup\{\tilde{V} : V \in \mathcal{V}(U)\}$$

and hence $U \subset \cup\{V : V \in \mathcal{V}(U)\}$. Finally we define

$$\mathcal{W} = \{U \cap V : U \in \mathcal{U},\ V \in \mathcal{V}(U)\}.$$

Clearly $\mathcal{W}$ is an open refinement of $\mathcal{V}$; and since $\mathcal{U}$ is locally finite in X and $\mathcal{V}(U)$ is finite for all $U \in \mathcal{U}$, the cover $\mathcal{W}$ is locally finite in X.

(d) Let X be topologically complete and $p \in \beta X \backslash X$. By Corollary 1.17 (with $\mathcal{P}$ the class of metric spaces) there are a metric space M and a continuous function $f : X \to M$ such that $\bar{f}(p) \in \beta M \backslash M$. Since M is paracompact (by Theorem 3.1) there is by (c) above a locally finite cozero cover $\mathcal{V}$ of M such that $V \in \mathcal{V}$ implies $\bar{f}(p) \notin c\ell_{\beta M} V$. We define

$$\mathcal{U} = \{f^{-1}(V) : V \in \mathcal{V}\}.$$

It is clear that $\mathcal{U}$ is a locally finite open cover of X, and that $V \in \mathcal{V}$ implies $p \notin cl_{\beta X} f^{-1}(V)$. Further if $V = \text{coz } g \in \mathcal{V}$ with $g \in C(M)$, then

$$f^{-1}(V) = \text{coz}(g \circ f) \text{ and } g \circ f \in C(X),$$

and hence $\mathcal{U}$ is a cozero cover.

For the converse let $p \in \beta X \backslash X$ and suppose that there is a locally finite cozero cover $\mathcal{U}$ of X such that $U \in \mathcal{U}$ implies $p \notin cl_{\beta X} U$. We set $\mathcal{U}_n = \mathcal{U}$ for $n < \omega$ and we define a metric space (M, ρ) and a (continuous) function $f : X \to M$ as in Theorem 3.2. Since $\bar{f}(p) \in \beta M \backslash M$ the space X is not M-embedded in $X \cup \{p\}$ and hence $p \notin \gamma X$ by Theorem 1.16. It follows that $\gamma X \backslash X = \emptyset$, so that X is topologically complete.

The proof is complete.

Remarks. We have noted already, in Corollary 2.4 and Lemma 4.2(a), that every Lindelöf space is both realcompact and paracompact. This is also obvious from Theorem 4.4, so that the following diagram of implications holds.

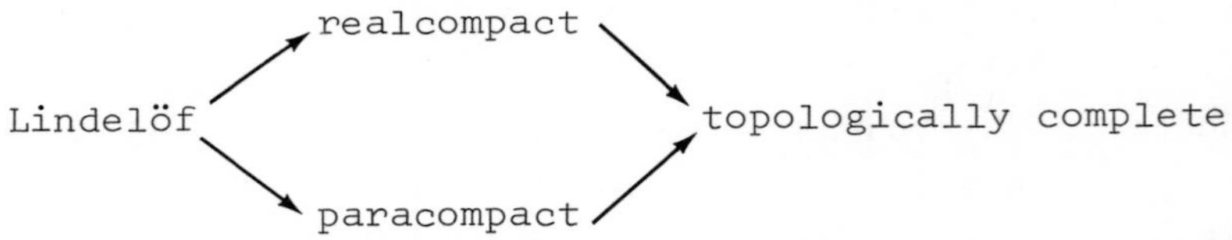

On the matter of other possible relationships between these four classes of spaces, we recall that the well-known "Sorgenfrey Line" is a Lindelöf space S such that $S \times S$ is not normal. Since the class of realcompact spaces is closed under the formation of products, $S \times S$ is by Lemma 4.2(b) a realcompact space that is not paracompact. As to the converse, the possibility that every paracompact space is realcompact is discussed in Corollary 6.4 and Remark 6.5 below.

Of the four statements in the following theorem, (a) and (c) can be proved directly without difficulty and (b) and (d) follow respectively from the fact that every compact space is realcompact and topologically complete; alternatively, the statements can be proved using elementary techniques from the theory of perfect functions. We choose instead to derive the results from the characterizations given in Theorem 4.4.

4.5. *Corollary.* Let X be a space and Y a compact space.

(a) If X is a Lindelöf space, then $X \times Y$ is a Lindelöf space;

(b) if X is realcompact, then $X \times Y$ is realcompact;

(c) if X is paracompact, then $X \times Y$ is paracompact;

(d) if X is topologically complete, then $X \times Y$ is topologically complete.

Proof. In order to apply Theorem 4.4, we let K be a compact subset of $\beta(X \times Y)\backslash(X \times Y)$ (with $|K| = 1$ in (b) and (d)); we prove that there is a locally finite cozero cover $\mathcal{U}$ of $X \times Y$ (with $|\mathcal{U}| \le \omega$ in (a) and (b)) such that $U \in \mathcal{U}$ implies $K \cap c\ell_{\beta(X\times Y)} U = \emptyset$.

Since $X \times Y \subset (\beta X) \times Y$ there is a continuous function

$$f : \beta(X \times Y) \to (\beta X) \times Y$$

such that $f(x,y) = \langle x,y\rangle$ for $\langle x,y\rangle \in X \times Y$; from Lemma 1.5 we have

$$f[K] \subset f[\beta(X \times Y)\backslash(X \times Y)] \subset (\beta X\backslash X) \times Y \subset (\beta X) \times Y.$$

We denote by π_0 the projection function from $(\beta X) \times Y$ onto βX and we set $\widetilde{K} = \pi_0[f[K]]$. Then $\widetilde{K}$ is a compact subset of $\beta X\backslash X$ and hence by Theorem 4.4 there is a locally finite cozero cover $\mathcal{V}$ of X (with $|\mathcal{V}| \le \omega$ in (a) and (b)) such that $V \in \mathcal{V}$ implies $\widetilde{K} \cap c\ell_{\beta X} V = \emptyset$. We set

$$\mathcal{U} = \{V \times Y : V \in \mathcal{V}\}.$$

Clearly $\mathcal{U}$ is a locally finite cozero cover of $X \times Y$ (with $|\mathcal{U}| \le \omega$ in (a) and (b)) and $V \times Y \in \mathcal{U}$ implies

$$f[K] \cap c\ell_{(\beta X)\times Y}(V \times Y) \subset (\widetilde{K} \times Y) \cap c\ell_{(\beta X)\times Y}(V \times Y) = \emptyset.$$

It follows that $V \times Y \in \mathcal{U}$ implies $K \cap c\ell_{\beta(X\times Y)}(V \times Y) = \emptyset$, since if $p \in K \cap c\ell_{\beta(X\times Y)}(V \times Y)$ then

$$f(p) \in f[K] \cap c\ell_{(\beta X)\times Y} f[V \times Y] = f[K] \cap c\ell_{(\beta X)\times Y}(V \times Y),$$

a contradiction.

§5. TAMANO'S THEOREM

We say that subsets A and B of a space X are *completely separated* (in X) if A and B are contained in disjoint zero-sets of X. Since $f^2/(f^2 + g^2)$ is a well-defined continuous function on X whenever $f, g \in C(X)$ and $Z(f) \cap Z(g) = \emptyset$, this condition is equivalent to the condition that there is $h \in C(X)$ such that $h[A] \subset \{0\}$ and $h[B] \subset \{1\}$. Urysohn's Lemma, a classical result (not proved here), states that disjoint closed subsets of a normal space are completely separated.

5.1. *Lemma.* Let X and Y be spaces with Y compact. Let $f \in C(X \times Y)$ and define $\phi : X \to R$ by

$$\phi(x) = \sup \{f(x,y) : y \in Y\}.$$

Then $\phi \in C(X)$.

Proof. Let $\bar{x} \in X$ and $\varepsilon > 0$, and for every $y \in Y$ let $U_y \times V_y$ be a neighborhood of $\langle\bar{x},y\rangle$ such that

$$|f(x',y') - f(\bar{x},y)| < \varepsilon/2$$

for $\langle x',y'\rangle \in U_y \times V_y$. There are $n < \omega$ and $\{y_k : k < n\} \subset Y$ such that $Y = \bigcup_{k<n} V_{y_k}$; we set $U = \bigcap_{k<n} U_{y_k}$ and we note that if $x \in U$ and $y \in Y$ (say $y \in V_{y_k}$) then

$$|f(x,y) - f(\bar{x},y)| \le |f(x,y) - f(\bar{x},y_k)| + |f(\bar{x},y_k) - f(\bar{x},y)|$$
$$< \varepsilon/2 + \varepsilon/2 = \varepsilon.$$

It follows that if $x \in U$ then $|\phi(x) - \phi(\bar{x})| < \varepsilon$. Indeed there are $x \in X$ and $\bar{y} \in Y$ such that

$$\phi(x) = f(x,y) \text{ and } \phi(\bar{x}) = f(\bar{x},\bar{y}),$$

and we have

$$\phi(\bar{x}) = f(\bar{x},\bar{y}) \ge f(\bar{x},y) > f(x,y) - \varepsilon = \phi(x) - \varepsilon, \text{ and}$$
$$\phi(x) = f(x,y) \ge f(x,\bar{y}) > f(\bar{x},\bar{y}) - \varepsilon = \phi(\bar{x}) - \varepsilon.$$

If Y is a space and (M,ρ) is a metric space, we denote by $C^*(Y,M)$ the set of all bounded continuous functions from Y into M, *i.e.*, all continuous $F : Y \to M$ such that $F[Y]$ has finite ρ-diameter. We define

$$\sigma : C^*(Y,M) \times C^*(Y,M) \to R$$

by the rule

$$\sigma(F,G) = \sup \{\rho(F(y),G(y)) : y \in Y\}$$

and we claim that σ is a metric for $C^*(Y,M)$. Indeed it is clear that if $F, G \in C^*(Y,M)$ then

$$0 \le \sigma(F,G) = \sigma(G,F), \text{ and}$$

$$\sigma(F,G) = 0 \text{ if and only if } F = G; \text{ and}$$

if $F, G, H \in C^*(Y,M)$ then

$$\begin{aligned} &\sup \{\rho(F(y),H(y)) : y \in Y\} \\ &\le \sup \{\rho(F(y),G(y)) + \rho(G(y),H(y)) : y \in Y\} \\ &\le \sup \{\rho(F(y),G(y)) : y \in Y\} \\ &+ \sup \{\rho(G(y),H(y)) : y \in Y\} \end{aligned}$$

and hence $\sigma(F,H) \le \sigma(F,G) + \sigma(G,H)$.

5.2. *Lemma*. Let X and Y be spaces with Y compact, let (M,ρ) be a metric space, let $f \in C^*(X \times Y, M)$, and for $x \in X$ define $F(x) : Y \to M$ by

$$F(x)(y) = f(x,y).$$

Then $F(x) \in C^*(Y,M)$, and (when $C^*(Y,M)$ has the topology defined by the metric

$$\sigma(G,H) = \sup \{\rho(G(y),H(y)) : y \in Y\})$$

the function $F : X \to C^*(Y,M)$ is continuous.

Proof. It is clear that $F(x) \in C^*(Y,M)$ for $x \in X$. We define $g : (X \times X) \times Y \to R$ by the rule

$$g(\langle x,x'\rangle,y) = \rho(f(x,y),f(x',y))$$

and for $\langle x,x'\rangle \in X \times X$ we define

$$\phi(x,x') = \sup \{g(\langle x,x'\rangle,y) : y \in Y\}.$$

Since $g \in C^*((X \times X) \times Y)$ and Y is compact we have $\phi \in C^*(X \times X)$ by Lemma 5.1. We note that

$$\phi(x,x') = \sigma(F(x),F(x')) \text{ for } \langle x,x'\rangle \in X \times X.$$

Now let $\bar{x} \in X$ and $\varepsilon > 0$. Since ϕ is continuous at $\langle\bar{x},\bar{x}\rangle$ there is a neighborhood N of $\bar{x}$ such that

$\phi(x,x') = |\phi(\bar{x},\bar{x}) - \phi(x,x')| < \varepsilon$ for $\langle x,x'\rangle \in N \times N$. In particular for every $x \in N$ we have

$$\sigma(F(x),F(\bar{x})) = \phi(x,\bar{x}) < \varepsilon;$$

it follows that the function F is continuous at $\bar{x}$.

We now prove Tamano's Theorem.

5.3. *Theorem.* For every space X, the following statements are equivalent.

(a) X is paracompact;

(b) $X \times Y$ is normal for every compact space Y;

(c) $X \times \beta X$ is normal;

(d) there is a compactification BX of X such that $X \times BX$ is normal.

Proof. (a) ⇒ (b) follows from Corollary 4.5(c) and Lemma 4.2(b); (b) ⇒ (c) and (c) ⇒ (d) are obvious.

To prove (d) ⇒ (a) it is (by Theorem 4.4(c)) enough to show that for every compact subset K of $\beta X \backslash X$ there is a locally finite cozero cover $\mathcal{U}$ of X such that $U \in \mathcal{U}$ implies $K \cap \mathrm{c\ell}_{\beta X} U = \emptyset$. There is a continuous function f from βX onto BX such that $f(x) = x$ for $x \in X$, and from Lemma 1.5 we have

$$f[K] \subset f[\beta X \backslash X] \subset BX \backslash X.$$

Clearly the two sets $\{\langle x,x\rangle : x \in X\}$ and $X \times f[K]$ are disjoint closed subsets of $X \times BX$. Hence since $X \times BX$ is normal there is $g \in C^*(X \times BX)$ such that

$$g(x,x) = 0 \text{ for } x \in X \text{ and } g[X \times f[K]] \subset \{1\}.$$

We define functions $h : X \times \beta X \to R$ and $H : X \to C^*(\beta X)$ by the rules

$$H(x)(p) = h(x,p) = g(x,f(p)) \text{ for } \langle x,p\rangle \in X \times \beta X$$

and we note that

$h \in C^*(X \times \beta X)$,
$h(x,x) = 0$ for all $x \in X$,
$h[X \times K] \subset \{1\}$, and (using Lemma 5.2)

H is a continuous function.

We set $M = H[X]$ and as usual for $H(x) \in M$ we define

$$S(H(x), 1/2) = \{q \in M : \rho(H(x), q) < 1/2\}.$$

Since (M,ρ) is paracompact by Theorem 3.1, the cover $\{S(H(x), 1/2) : x \in X\}$ has a locally finite cozero refinement $\mathcal{V}$. We set

$$\mathcal{U} = \{H^{-1}(V) : V \in \mathcal{V}\}$$

and we note that $\mathcal{U}$ is a locally finite cozero cover of X. We claim also that

$$U = H^{-1}(V) \in \mathcal{U} \text{ implies } K \cap c\ell_{\beta X} U = \emptyset.$$

Indeed let $p \in c\ell_{\beta X} U$ and let $\bar{x}$ be an element of X such that $V \subset S(H(\bar{x}), 1/2)$. For every neighborhood W of p in βX there is $\bar{y} \in W \cap U$, and since

$$H(\bar{y}) \in V \subset S(H(\bar{x}), 1/2)$$

we have $|h(\bar{x},\bar{y})| = |h(\bar{x},\bar{y}) - h(\bar{y},\bar{y})| < 1/2$. Thus every neighborhood in $X \times \beta X$ of $(\bar{x},p)$ contains a point $(\bar{x},\bar{y})$ such that $h(\bar{x},\bar{y}) < 1/2$; hence $h(\bar{x},p) \le 1/2$. Since $h[X \times K] \subset \{1\}$ we have $p \notin K$, as required. The proof is complete.

We note (from the proof of Tamano's theorem) that the special spaces $X \times \beta X$ are normal if and only if for every compact subset K of $\beta X \setminus X$ the disjoint sets $\{\langle x,x \rangle : x \in X\}$ and $X \times K$ are completely separated in $X \times \beta X$.

We conclude this Section with a partial analogue of Theorem 5.3 for topologically complete spaces. The implication (a) ⇒ (b) below is used for one of the proofs of the theorem of Katětov-Shirota.

5.4. *Theorem.* For every space X, the following statements are equivalent.

(a) X is topologically complete;

(b) for every $p \in \beta X \setminus X$ the sets $X \times \{p\}$ and $\{\langle x,x \rangle : x \in X\}$ are completely separated in $X \times \beta X$.

Proof. (a) ⇒ (b). Since X is topologically complete

and $p \in \beta X \backslash X$ there are by Theorem 1.16 a metric space M and a continuous function $f : \beta X \to \beta M$ such that $f[X] \subset M$ and $f(p) \in \beta M \backslash M$. We define

$$X' = f^{-1}(M)$$

and we note from Lemma 1.9 that X' is homeomorphic to a closed subspace of $\beta X \times M$. Since M is paracompact (Theorem 3.1(a)) the space $\beta X \times M$ is paracompact by Corollary 4.5(c). It follows that X' is paracompact, and hence $X' \times \beta X'$ is normal by Tamano's theorem and there is $g \in C^*(X' \times \beta X')$ such that

$$g[X' \times \{p\}] \subset \{1\} \quad \text{and} \quad g(x,x) = 0 \quad \text{for} \quad x \in X'.$$

We note that $X \subset X' \subset \beta X$, so that $\beta X' = \beta X$ by Lemma 1.6. We set $h = g|X \times \beta X$, and we have

$$h[X \times \{p\}] \subset \{1\} \quad \text{and} \quad h(x,x) = 0 \quad \text{for} \quad x \in X,$$

as required.

(b) ⇒ (a). For $p \in \beta X \backslash X$ there is a continuous function $h : X \times \beta X \to [0,1]$ such that

$$h(x,x) = 0 \quad \text{for all} \quad x \in X \quad \text{and} \quad h[X \times \{p\}] \subset \{1\};$$

as in the proof that (d) ⇒ (a) in Theorem 5.3 we define $H : X \to C^*(\beta X)$ by $H(x)(q) = h(x,q)$, we define $M = H[X]$ and choose a locally finite cozero refinement $\mathcal{V}$ of $\{S(H(x),1/2) : x \in X\}$, and we set $\mathcal{U} = \{H^{-1}(V) : V \in \mathcal{V}\}$. The proof that $p \notin \cup\{c\ell_{\beta X}\, U : U \in \mathcal{U}\}$ procedes then as in Theorem 5.3((d) ⇒ (a)). We omit the details.

§6. THE KATĚTOV-SHIROTA THEOREM

We have seen in Theorem 2.6 that for every cardinal number α, the discrete space α is realcompact if and only if α is not Ulam-measurable. It is unreasonable to conjecture that, more generally, a topologically complete space should be realcompact if and only if its cardinal number is not Ulam-measurable: If α is an Ulam-measurable cardinal and $X = R^{\alpha}$ or $X = \{0,1\}^{\alpha}$, for example, then X is realcompact but according to Lemma 2.5(a) the cardinal $|X|$ is Ulam-measurable. The Katětov-Shirota Theorem (6.3 below) determines whether or not a topologically complete space is realcompact on the basis of whether or not it has a closed, discrete subset of Ulam-measurable cardinal. The transition between discrete spaces and arbitrary topologically complete spaces is made *via* metric spaces. According to Theorem 6.2 a metric space M is not realcompact if and only if $|M|$ is Ulam-measurable, and if these (equivalent) conditions are satisfied then M has a closed, discrete subset of Ulam-measurable cardinal.

6.1. *Lemma.* If X is a realcompact space and A is a closed, discrete subset of X, then $|A|$ is not Ulam-measurable.

Proof. Since every closed subspace of a realcompact space is realcompact, this follows from Theorem 2.6.

6.2. *Theorem.* For every metric space (M,ρ), the following statements are equivalent.

(a) M is realcompact;

(b) $|M|$ is not Ulam-measurable;

(c) for every closed, discrete subset A of M, $|A|$ is not Ulam-measurable.

Proof. (a) ⇒ (b). For $1 \leq n < \omega$ there is by Zorn's lemma a subset A_n of M such that

(i) if $x, y \in A_n$ and $x \neq y$ then $\rho(x,y) \geq 1/n$, and

(ii) if $A_n \subset B \subset M$ and if $\rho(x,y) \geq 1/n$ for $x, y \in B$ with $x \neq y$, then $B = A_n$.

We define $A = \cup_{1 \leq n < \omega} A_n$ and we note that A is dense in M.

Since A_n is closed and discrete we have from Lemma 6.1 that $|A_n|$ is not Ulam-measurable, and hence $|A|$ is not Ulam-measurable (by Lemma 2.5(b)). From Lemma 1.1 we have $|M| \leq 2^{2^{|A|}}$, so that $|M|$ is not Ulam-measurable (by Lemma 2.5(c)).

(b) ⇒ (c) follows from Lemma 2.5(a).

(c) ⇒ (a). To show that M is realcompact it is sufficient (according to Theorem 2.2) to prove that for every $p \in \beta M \backslash M$ there is $Z \in \mathcal{Z}(\beta M)$ such that $p \in Z \subset \beta M \backslash M$. We fix $p \in \beta M \backslash M$.

For every $x \in M$ there is an open neighborhood N_x of x in M such that $p \notin c\ell_{\beta M} N_x$, and by Theorem 3.1(b) the cover $\{N_x : x \in M\}$ has an open refinement $\mathcal{U} = \cup_{n<\omega} \mathcal{U}_n$ with $\mathcal{U}_n$ discrete in M for $n < \omega$. We define

$$U_n = \cup \mathcal{U}_n \text{ for } n < \omega$$

and we note that $\cup_{n<\omega} U_n = M$. We consider two cases.

Case 1. $p \notin \cup_{n<\omega} c\ell_{\beta M} U_n$.
For $n < \omega$ there is a continuous function

$$f_n : \beta M \to [0,1]$$

such that $f_n(p) = 0$ and $f_n[U_n] \subset \{1\}$. We define

$$f = \Sigma_{n<\omega} f_n/2^n$$

and we note that $f \in C(\beta M)$ and that

$$p \in Z(f) \subset \beta M \backslash M.$$

The proof is complete (under the assumption of Case 1).

Case 2. There is $n < \omega$ such that $p \in c\ell_{\beta M} U_n$. Let $\{V_i : i \in I\}$ be a faithful indexing of the (non-empty) elements of $\mathcal{U}_n$, and for $i \in I$ choose $x_i \in V_i$. The function $i \to x_i$ is a one-to-one function from I onto a closed, discrete subset of M, and hence $|I|$ is not an Ulam-measurable cardinal. We define a function

$$\mu : \mathcal{P}(I) \to \{0,1\}$$

by the rule

$$\mu(J) = 1 \text{ if } p \in c\ell_{\beta M}(\cup_{i \in J} V_i)$$
$$= 0 \text{ if } p \notin c\ell_{\beta M}(\cup_{i \in J} V_i).$$

It is clear that $\mu(I) = 1$; further for every $i \in I$ there is $x \in M$ such that $V_i \subset N_x$, and since $p \notin c\ell_{\beta M} N_x$ we have $p \notin c\ell_{\beta X} V_i$ and hence $\mu(\{i\}) = 0$ for $i \in I$. We claim next that if $J \subset I$ then

$$\mu(J) = 0 \text{ if and only if } \mu(I \backslash J) = 1.$$

Indeed since $\{V_i : i \in I\}$ is a discrete family the sets

$$\cup_{i \in J} V_i \text{ and } \cup_{i \in I \backslash J} V_i$$

have disjoint closures in the (normal) space M and hence there is a continuous function $f : M \to [0,1]$ such that

$$f[\cup_{i \in J} V_i] \subset \{0\} \text{ and } f[\cup_{i \in I \backslash J} V_i] \subset \{1\};$$

it is then clear (since $p \in c\ell_{\beta M} U_n$) that p is in the closure in βM of exactly one of the two sets $\cup_{i \in J} V_i$, $\cup_{i \in I \backslash J} V_i$. The claim is proved, and it follows that μ is a finitely additive $\{0,1\}$-valued measure. Hence since $|I|$ is not Ulam-measurable there is $\{J_k : k < \omega\} \subset \mathcal{P}(I)$ such that

$$\mu(J_k) = 0 \text{ for } k < \omega \text{ and}$$
$$\mu(\cup_{k<\omega} J_k) = 1.$$

We define

$$J = \cup_{k<\omega} J_k,$$
$$B_k = \cup\{V_i : i \in J_k\} \text{ for } k < \omega, \text{ and}$$
$$B = \cup_{k<\omega} B_k.$$

Since $c\ell_M B$ is a closed subset of M there is a continuous function $g : M \to [0,1]$ such that $Z(g) = c\ell_M B$. (Specifically, one may define g by

$$g(x) = \min \{1, \rho(x, c\ell_M B)\} \text{ for } x \in M.)$$

We note that $\bar{g}(p) = 0$ since $p \in c\ell_{\beta M} B$; and we note also, using the fact that

$$c\ell_M B = c\ell_M(\cup_{i\in J} V_i) = \cup_{i\in J} c\ell_M V_i,$$

that $g(x) > 0$ for all $x \in M\backslash(\cup_{i\in J} c\ell_M V_i)$.

Since $p \notin c\ell_{\beta M} B_k$ for $k < \omega$ there is a continuous function $h_k : \beta M \to [0,1]$ such that

$$h_k(p) = 0 \text{ and } h_k[B_k] \subset \{1\}.$$

We define

$$h = \bar{g} + \Sigma_{k<\omega} h_k/2^k$$

and we note that $p \in Z(h)$. Further if $x \in M\backslash(\cup_{i\in J} c\ell_M V_i)$ then $h(x) \geq \bar{g}(x) = g(x) > 0$, and if $x \in \cup_{i\in J} c\ell_M V_i$ then there are $k < \omega$ and $i \in J_k$ such that

$$x \in c\ell_M V_i \subset c\ell_M B_k$$

and $h(x) \geq 1/2^k > 0$. It follows that $Z(h) \subset \beta M\backslash M$. The proof is complete.

Here is the Katětov-Shirota theorem.

6.3. *Theorem.* For every topologically complete space X, the following are equivalent.

(a) X is realcompact;

(b) for every closed, discrete subset A of X, $|A|$ is not Ulam-measurable.

Proof. (a) $\Rightarrow$ (b) follows from Lemma 6.1.

(b) $\Rightarrow$ (a). We give two proofs of this implication. We remark (for use in each proof) that if f is a continuous function from X onto a space Y, then every closed, discrete subset of Y is of non-Ulam-measurable cardinal. Indeed if D is a closed, discrete subset of Y and for every $y \in Y$ we choose $x(y) \in f^{-1}(\{y\})$, then $\{x(y) : y \in Y\}$ is a closed, discrete subset of X and hence is of non-Ulam-measurable cardinal.

First Proof. Since X is topologically complete, for every $p \in \beta X \backslash X$ there are by Theorem 2.1 a metric space M_p and a continuous function f_p from X onto M_p such that $\overline{f_p}(p) \in \beta(M_p) \backslash M_p$. It follows from Theorem 6.2 and the remark above that M_p is realcompact for $p \in \beta X \backslash X$; hence (from Corollary 1.10, with P the class of realcompact spaces), the space $\overline{f_p}^{\,-1}(M_p)$ is realcompact for $p \in \beta X \backslash X$. Since

$$X = \cap\{\overline{f_p}^{\,-1}(M_p) : p \in \beta X \backslash X\},$$

it follows from Lemma 1.8 that X is realcompact.

Second Proof. We prove that if $p \in \beta X \backslash X$ then $p \notin \upsilon X$. For $p \in \beta X \backslash X$ there is by Theorem 5.4 a continuous function $f : X \times \beta X \to [0,1]$ such that

$$f(x,x) = 0 \quad \text{for} \quad x \in X \quad \text{and} \quad f[X \times \{p\}] \subset \{1\}.$$

For $x \in X$ and $q \in \beta X$ we define

$$F(x)(q) = f(x,q)$$

and we note from Lemma 5.2 that $F(x) \in C^*(\beta X)$ for $x \in X$ and that $F : X \to C^*(\beta X)$ is a continuous function. By Theorem 6.2 and the remark preceding the First Proof, the metric space $F[X]$ is realcompact and hence there is a continuous function

$$\tilde{F} : \upsilon X \to F[X] \subset C^*(\beta X)$$

such that $\tilde{F}|X = F$. We define $\tilde{f} : \upsilon X \times \beta X \to R$ by

$$\tilde{f}(x,q) = \tilde{F}(x)(q) \quad \text{for} \quad \langle x,q\rangle \in \upsilon X \times \beta X$$

and we claim that $\tilde{f} \in C^*(\upsilon X \times \beta X)$. Indeed if $\langle \bar{x},\bar{q}\rangle \in \upsilon X \times \beta X$ and $\varepsilon > 0$ then there is a neighborhood U of $\bar{x}$ in υX such that

$$\rho(\tilde{F}(x),\tilde{F}(\bar{x})) < \varepsilon/2 \quad \text{for} \quad x \in U$$

and there is a neighborhood V of $\bar{q}$ in βX such that

$$|\tilde{F}(\bar{x})(q) - \tilde{F}(\bar{x})(\bar{q})| < \varepsilon/2 \quad \text{for} \quad q \in V;$$

then for $\langle x,q\rangle \in U \times V$ we have

$$|\tilde{f}(x,q) - \tilde{f}(\bar{x},\bar{q})| \le |\tilde{f}(x,q) - \tilde{f}(\bar{x},q)| + |\tilde{f}(\bar{x},q) - \tilde{f}(\bar{x},\bar{q})| < \varepsilon/2 + \varepsilon/2 = \varepsilon.$$

Thus $\tilde{f}$ is a continuous function. It is clear that $f \subset \tilde{f}$. Suppose, finally, that $p \in \upsilon X$. Since

$$\langle p,p\rangle \in c\ell_{\upsilon X \times \beta X}\, \{\langle x,x\rangle : x \in X\} \cap c\ell_{\upsilon X \times \beta X}(X \times \{p\})$$

we have $\langle p,p\rangle \in c\ell_{\upsilon X \times \beta X}\, \tilde{f}^{-1}(\{0\}) \cap c\ell_{\upsilon X \times \beta X}\, \tilde{f}^{-1}(\{1\})$ and hence

$$\tilde{f}(p,p) = 0 \quad \text{and} \quad \tilde{f}(p,p) = 1.$$

This contradiction completes the (second) proof.

6.4. *Corollary.* The following statements are equivalent.

(a) There is no Ulam-measurable cardinal;

(b) every topologically complete space is realcompact;

(c) every paracompact space is realcompact;

(d) every metric space is realcompact;

(e) every discrete space is realcompact.

Proof. The implications (e) ⇒ (a) ⇒ (b) ⇒ (c) ⇒ (d) are given by Theorems 2.6, 6.3, 4.4 and 3.1 respectively, and (d) ⇒ (e) is obvious.

6.5. *Remark.* We denote by ZFC the usual Zemelo-Fraenkel Set Theory, together with the Axiom of Choice. This is the setting for these Notes, it being assumed that ZFC is a consistent theory, *i.e.*, that there is a model for ZFC.

It is not known whether the axiom

(U) There is an Ulam-measurable cardinal

can be adjoined to ZFC without introducing any contradiction.

We note, however, that (U) is not a theorem in ZFC. Indeed if α is the smallest Ulam-measurable cardinal in a model M of ZFC then the class of sets of M whose rank in M is less than α is a submodel of M of ZFC in which (U) fails.

Thus the systems ZFC and ZFC + ¬(U) are equiconsistent.

§7. On the Relations $P(X \times Y) = PX \times PY$

We consider here pairs of spaces $<X,Y>$ such that $\gamma(X \times Y) = \gamma X \times \gamma Y$, and similarly for υ and β. In general the results are less definitive than those on other topics presented elsewhere in these Notes; we include this Section in the hope that it will stimulate some readers to consider certain unsolved problems, and to complete parts of the theory.

We begin with an example showing that the relation $\upsilon(X \times Y) = \upsilon X \times \upsilon Y$ can fail (and simultaneously for β and γ). Let ω denote the (discrete) space of finite cardinals. It is well-known (see for example Gillman and Jerison [60] (Theorem 9.2) or Comfort and Negrepontis [74] (Corollary 7.4)) that $|\beta(\omega)| = 2^{2^\omega}$; it follows that $|c\ell_{\beta(\omega)} A| = 2^{2^\omega}$ for every infinite subset A of ω. Let $<A_\xi : \xi < 2^\omega>$ be a well-ordering of the set of infinite subsets of ω, let p_0 and q_0 be distinct elements of $(c\ell_{\beta(\omega)} A_0)\backslash\omega$, and recursively, if $\xi < 2^\omega$ and if p_ζ and q_ζ have been chosen for $\zeta < \xi$, let p_ξ and q_ξ be distinct elements of $(c\ell_{\beta(\omega)} A_\xi)\backslash\omega$ such that

$$p_\xi, q_\xi \notin \{p_\zeta : \zeta < \xi\} \cup \{q_\zeta : \zeta < \xi\}.$$

Define $P = \{p_\xi : \xi < 2^\omega\}$ and $Q = \{q_\xi : \xi < 2^\omega\}$, and set

$$X = \omega \cup P \quad \text{and} \quad Y = \omega \cup Q.$$

Since $P \cap Q = \emptyset$ the set

$$\Delta = \{<n,n> : n < \omega\}$$

is open-and-closed in $X \times Y$. We define

$$f : X \times Y \to [0,1]$$

by the rule

$$f[\Delta] \subset \{0\},$$
$$f[(X \times Y)\backslash\Delta] \subset \{1\}$$

and we note that $f \in C^*(X \times Y)$. We note also that if $p \in \beta(\omega)\backslash\omega$ then

$$\langle p,p\rangle \in cl_{\beta(\omega)\times\beta(\omega)}\, f^{-1}(\{0\}) \cap cl_{\beta(\omega)\times\beta(\omega)}\, f^{-1}(\{1\})$$

and hence there is no continuous function

$$g : \beta(\omega)\times\beta(\omega) \to [0,1]$$

such that $f \subset g$. Since $\beta X = \beta Y = \beta(\omega)$, it follows that $\beta(X \times Y) \neq \beta X \times \beta Y$. We show now that $\upsilon(X \times Y) \neq \upsilon X \times \upsilon Y$ and $\gamma(X \times Y) \neq \gamma X \times \gamma Y$.

We claim that $\upsilon X = \beta(\omega)$. If there is $p \in \beta(\omega)\backslash\upsilon X = \beta X\backslash\upsilon X$ then by Theorem 2.2 there is a continuous function $g : \beta X \to [0,1]$ such that

$$p \in Z(g) \subset \beta X\backslash\upsilon X,$$

and since $p \in cl_{\beta(\omega)}\, \omega$ there is $\{n_k : k < \omega\} \subset \omega$ such that $g(n_k) < 1/k$. Further there is $\xi < 2^{\omega}$ such that

$$A_\xi = \{n_k : k < \omega\},$$

and since $g[(cl_{\beta(\omega)}\, A_\xi)\backslash\omega] \subset \{0\}$ we have $g(p_\xi) = 0$. Thus $p_\xi \in \beta(\omega)\backslash\upsilon X$, a contradiction. The claim is proved; a similar argument shows that $\upsilon Y = \beta(\omega)$.

Suppose now that $\upsilon(X \times Y) = \upsilon X \times \upsilon Y$. Then $\upsilon(X \times Y) = \beta X \times \beta Y$ and hence $X \times Y$ is C-embedded in $\beta X \times \beta Y$. It follows that $X \times Y$ is C*-embedded in $\beta X \times \beta Y$, so that $\beta(X \times Y) = \beta X \times \beta Y$, a contradiction. We conclude that $\upsilon(X \times Y) \neq \upsilon X \times \upsilon Y$.

Finally we note that if Z is one of the spaces X, Y or $X \times Y$ then Z (and hence υZ) is separable and hence $|\upsilon Z| \leq 2^{2^{\omega}}$ by Lemma 1.1. It follows from Lemma 2.5(a),(c) and Theorem 6.3 that γZ is realcompact and hence $\gamma Z = \upsilon Z$; thus we have

$$\gamma(X \times Y) \neq \gamma X \times \gamma Y,$$

as required. The proof is complete.

We proceed now to determine conditions on $\langle X,Y\rangle$ such that $\upsilon(X \times Y) = \upsilon X \times \upsilon Y$ or $\gamma(X \times Y) = \gamma X \times \gamma Y$.

7.1. *Lemma.* Let P be a class of spaces such that $C \subset P = \bar{P}$ and let X and Y be spaces such that $X \times Y$

is C*-embedded in $PX \times PY$. Then $X \times Y$ is P-embedded in $PX \times PY$.

Proof. We note from Theorem 1.13 that $X \times Y$ is C-embedded in $PX \times PY$.

Given a continuous function $f : X \times Y \to Z \in P$ there is a continuous function $\tilde{f} : PX \times PY \to \beta Z$ such that $f \subset \tilde{f}$. It is enough to show that $\tilde{f}[PX \times PY] \subset Z$. We show first that $\tilde{f}[X \times PY] \subset Z$.

For $x \in X$ we define $f_x = f|\{x\} \times Y$ and we denote by $\tilde{f}_x$ the continuous function from $\{x\} \times PY$ to Z such that $f_x \subset \tilde{f}_x$. Since $\tilde{f}|\{x\} \times Y = \tilde{f}_x|\{x\} \times Y$ and $\tilde{f}|\{x\} \times PY$ and $\tilde{f}_x$ are continuous on $\{x\} \times PY$, it follows from Lemma 1.4 that $\tilde{f}|\{x\} \times PY = \tilde{f}_x$. Hence

$$\tilde{f}[X \times PY] = \cup_{x\in X} \tilde{f}[\{x\} \times PY] = \cup_{x\in X} \tilde{f}_x[\{x\} \times PY] \subset \cup_{x\in X} Z = Z.$$

A similar argument now shows that $\tilde{f}[PX \times PY] \subset Z$. The proof is complete.

7.2. *Theorem.* Let P and Q be classes of spaces such that

$$C \subset Q = \overline{Q} \subset P = \overline{P}$$

and let X and Y be spaces such that $Q(X \times Y) = QX \times QY$. Then $P(X \times Y) = PX \times PY$.

Proof. We show that $X \times Y$ is P-embedded in $PX \times PY$. Since $X \times Y$ is Q-embedded in $QX \times QY$, and since (by Corollary 1.12)

$$PX \subset QX \subset \beta X \quad \text{and} \quad PY \subset QY \subset \beta Y,$$

clearly $X \times Y$ is Q-embedded in $PX \times PY$. Hence $X \times Y$ is C*-embedded in $PX \times PY$ and the result follows from Lemma 7.1.

7.3. *Lemma.* Let X be a space and Y a compact space. Then

(a) $X \times Y$ is M-embedded in $(\gamma X) \times Y$; and

(b) if $|X|$ or $|Y|$ is not Ulam-measurable, then $X \times Y$ is C^*-embedded in $(\upsilon X) \times Y$.

Proof. In the proof of (a) we set $\tilde{X} = \gamma X$, and in the proof of (b) we set $\tilde{X} = \upsilon X$. Let f be a continuous function from $X \times Y$ to a metric space (M,ρ), with ρ chosen in (a) so that M has finite ρ-diameter, and with $M = [0,1]$ in (b). We show that there is a continuous function $\tilde{f} : \tilde{X} \times Y \to M$ such that $f \subset \tilde{f}$. For $x \in X$ we define $F(x) : Y \to M$ by

$$F(x)(y) = f(x,y) \quad \text{for} \quad y \in Y$$

and we note from Lemma 5.2 that the function

$$F : X \to C^*(Y,M)$$

is continuous. Under the conditions of (b) we have

$$|F[X]| \le |X| \quad \text{and} \quad |F[X]| \le |R^Y| = 2^{\omega \cdot |Y|}$$

and hence $F[X]$ is a realcompact metric space by Lemma 2.5 and Theorem 6.3. Thus in both (a) and (b) there is a continuous function $\tilde{F} : \tilde{X} \to C^*(Y,M)$ such that $F \subset \tilde{F}$. We define $\tilde{f} : \tilde{X} \times Y \to M$ by

$$\tilde{f}(p,y) = \tilde{F}(p)(y) \quad \text{for} \quad \langle p,y\rangle \in \tilde{X} \times Y,$$

we note that $f \subset \tilde{f}$, and we claim that $\tilde{f} \in C^*(\tilde{X} \times Y, M)$. Indeed if $\langle \bar{p},\bar{y}\rangle \in \tilde{X} \times Y$ and $\varepsilon > 0$ there are neighborhoods U and V of $\bar{p}$ and $\bar{y}$ respectively such that

$$\sigma(\tilde{F}(p),\tilde{F}(\bar{p})) < \varepsilon/2 \quad \text{for} \quad p \in U \quad \text{and}$$

$$\rho(\tilde{F}(\bar{p})(y),\tilde{F}(\bar{p})(\bar{y})) < \varepsilon/2 \quad \text{for} \quad y \in V,$$

and for $\langle p,y\rangle \in U \times V$ we have

$$\rho(\tilde{f}(p,y),\tilde{f}(\bar{p},\bar{y})) \le \rho(\tilde{f}(p,y),\tilde{f}(\bar{p},y)) + \rho(\tilde{f}(\bar{p},y),\tilde{f}(\bar{p},\bar{y}))$$
$$< \varepsilon/2 + \varepsilon/2 = \varepsilon.$$

Thus $\tilde{f}$ is continuous. The proof is complete.

7.4. *Corollary*. Let X be a space and Y a locally compact space.

(a) If Y is topologically complete then $\gamma(X \times Y) = \gamma X \times Y$;

(b) if Y is realcompact and either $|X|$ or $|Y|$ is not Ulam-measurable, then $\upsilon(X \times Y) = \upsilon X \times \upsilon Y$.

Proof. (a) It follows from Lemma 7.3(a) that for every continuous function $f : X \times Y \to M$ with M metrizable and for every $y \in Y$ there are a compact neighborhood K_y of y and a continuous function

$$\tilde{f}_y : (\gamma X) \times K_y \to M$$

such that $\tilde{f}_y | X \times K_y = f | X \times K_y$. It is clear that if $y, y' \in Y$ and $y \in K_{y'}$, then

$$\tilde{f}_y(p,y) = \tilde{f}_{y'}(p,y) \quad \text{for} \quad p \in \gamma X;$$

hence $\tilde{f} = \cup_{y \in Y} \tilde{f}_y$ is a well-defined continuous function on $(\gamma X) \times Y$. It follows that $X \times Y$ is M-embedded in $(\gamma X) \times Y$, as required.

(b) An argument similar to that of part (a), using Lemma 7.3(b) in place of 7.3(a), proves that $X \times Y$ is C^*-embedded in $(\upsilon X) \times Y$. The required conclusion then follows from Lemma 7.1.

We conclude this Section with Glicksberg's Theorem, which is the one satisfying result concerning relations of the form $P(X \times Y) = PX \times PY$.

Definition. A space X is *pseudocompact* if every real-valued continuous function on X is bounded, *i.e.*, if $C(X) = C^*(X)$.

It is clear that a space X is pseudocompact if and only if for every $f \in C^*(X)$ such that

$$\inf \{f(x) : x \in X\} = 0,$$

the set $Z(f)$ is non-empty.

7.5. *Lemma.* Let X be a space. The following statements are equivalent.

(a) X is pseudocompact;

(b) every locally finite family of non-empty open subsets of X is finite;

(c) every discrete family of non-empty open subsets of X is finite.

Proof. (a) ⇒ (b). Let $\{U_n : n < \omega\}$ be a locally finite family of non-empty subsets of X, and for $n < \omega$ let $x_n \in U_n$ and let f_n be a continuous function such that

$$f_n(x_n) = n,$$
$$f_n(x) = 0 \quad \text{for} \quad x \in X \setminus U_n, \quad \text{and}$$
$$f_n(x) \geq 0 \quad \text{for} \quad x \in X.$$

We define $f = \Sigma_{n<\omega} f_n$. Since $\{U_n : n < \omega\}$ is a locally finite family, the function f is a well-defined, real-valued continuous function on X. Clearly $f \notin C^*(X)$, and hence X is not pseudocompact.

(b) ⇒ (c) is clear.

(c) ⇒ (a). If (a) fails there is $f \in C(X)\setminus C^*(X)$ such that $f(x) \geq 0$ for $x \in X$. We choose $x_0 \in X$, recursively for $n < \omega$ we choose $x_n \in X$ such that

$$f(x_n) > f(x_{n-1}) + 1,$$

and for $n < \omega$ we set $U_n = \{x \in X : |f(x) - f(x_n)| < 1/3\}$. Then $\{U_n : n < \omega\}$ is an infinite, discrete family of non-empty open subsets of X.

We note that from Lemma 7.5 it follows that every pseudocompact, topologically complete space X is compact. Indeed if there is $p \in \beta X \setminus X$ then by Theorem 4.4(d) there is a locally finite cozero cover $\mathcal{U}$ of X such that $U \in \mathcal{U}$ implies $p \notin c\ell_{\beta X}\, U$; since $\mathcal{U}$ is finite (by Lemma 7.5(a) ⇒ (b)) we have

$$p \notin \cup\{c\ell_{\beta X}\, U : U \in \mathcal{U}\} = c\ell_{\beta X}(\cup\, \mathcal{U}) = c\ell_{\beta X}\, X = \beta X,$$

a contradiction. It follows in particular that every pseudocompact metric space is compact.

7.6. *Theorem*. For every two spaces X and Y, the following statements are equivalent.

(a) $\beta(X \times Y) = \beta X \times \beta Y$;

(b) either X or Y is finite, or $X \times Y$ is pseudocompact.

Proof. (a) ⇒ (b). We assume that neither X nor Y

is finite. We claim first that X is pseudocompact. If the claim fails there is by Lemma 7.5 a discrete family $\{U_n : n < \omega\}$ of non-empty open subsets of X. We choose $x_n \in U_n$ for $n < \omega$. We choose a sequence $\{y_n : n < \omega\}$ of distinct elements of Y and we choose $q \in \beta Y$ such that if V is a neighborhood in βY of q then $|\{n < \omega : y_n \in V\}| = \omega$; we assume without loss of generality that if $n < \omega$ then $q \neq y_n$.

It is clear that if $n < \omega$ then $q \notin c\ell_{\beta Y}\{y_k : k \leq n\}$; hence there is a continuous function $f_n : X \times \beta Y \to [0,1]$ such that

$$f_n(x_n, y_k) = 0 \quad \text{for} \quad k \leq n,$$
$$f_n(x_n, q) = 1, \quad \text{and}$$
$$f_n[(X \backslash U_n) \times \beta Y] \subset \{0\}.$$

Since $\{\text{coz } f_n : n < \omega\}$ is a discrete family in $X \times \beta Y$, the function

$$f = \Sigma_{n<\omega} f_n$$

is a well-defined continuous function from $X \times \beta Y$ to $[0,1]$, and hence from (a) there is $\tilde{f} \in C^*(\beta X \times \beta Y)$ such that $f \subset \tilde{f}$. There is $p \in \beta X$ such that if U is a neighborhood in βX of p then $|\{n < \omega : x_n \in U\}| = \omega$. We note that for every $k < \omega$ we have

$$f(x_n, y_k) = f_n(x_n, y_k) = 0 \quad \text{for} \quad n \geq k$$

and hence $\tilde{f}(p, y_k) = 0$; it follows that $\tilde{f}(p,q) = 0$. But

$$\langle p,q \rangle \in c\ell_{\beta X \times \beta Y}\{\langle x_n, q \rangle : n < \omega\}$$

and $f(x_n, q) = f_n(x_n, q) = 1$ for $n < \omega$, so that $\tilde{f}(p,q) = 1$. This contradiction completes the proof that X is pseudocompact. A similar argument proves that Y is pseudocompact.

To prove, finally, that $X \times Y$ is pseudocompact it is enough to verify that if $f \in C^*(X \times Y)$ and

$$\inf \{f(x,y) : \langle x,y \rangle \in X \times Y\} = 0,$$

then $Z(f) \neq \emptyset$. There is $\tilde{f} \in \beta X \times \beta Y$ such that $f \subset \tilde{f}$, and since $\beta X \times \beta Y$ is compact there is $\langle p,q \rangle \in Z(\tilde{f})$. Since $\tilde{f}|\{p\} \times Y$ is continuous on the pseudocompact space $\{p\} \times Y$ and

$$\inf \{\tilde{f}(p,y) : y \in Y\} = 0,$$

there is $\bar{y} \in Y$ such that $\langle p,\bar{y}\rangle \in Z(\tilde{f})$. Similarly since $\tilde{f}|X \times \{\bar{y}\}$ is continuous there is $\bar{x} \in X$ such that

$$\langle\bar{x},\bar{y}\rangle \in Z(\tilde{f}) \cap X \times Y = Z(f).$$

(b) ⇒ (a). It is clear that if X or Y is finite then $X \times Y$ is C*-embedded in $\beta X \times \beta Y$ and hence $\beta(X \times Y) = \beta X \times \beta Y$. We assume that $X \times Y$ is pseudocompact and we claim first that $X \times Y$ is C*-embedded in $X \times \beta Y$. Let $f \in C^*(X \times Y)$, for $x \in X$ define $f_x \in C^*(Y)$ by

$$f_x(y) = f(x,y) \quad \text{for } y \in Y,$$

let $f_x \subset \tilde{f}_x \in C^*(\beta Y)$ and define

$$\tilde{f} = \cup_{x\in X} \tilde{f}_x.$$

If $\tilde{f}$ is not continuous on $X \times \beta Y$ then by Lemma 1.15 there are $\langle\bar{x},\bar{q}\rangle \in X \times \beta Y$ and $\varepsilon > 0$ such that for every neighborhood N of $\langle\bar{x},\bar{q}\rangle$ there is $\langle x,y\rangle \in N \cap (X \times Y)$ such that

$$|f(x,y) - \tilde{f}(\bar{x},\bar{q})| \geq \varepsilon.$$

We assume without loss of generality, replacing if necessary the function f by the function

$$\langle x,y\rangle \to 1 - \min\{1, |f(x,y) - \tilde{f}(\bar{x},\bar{q})|/\varepsilon\},$$

that

$$\tilde{f}(\bar{x},\bar{q}) = 1, \quad \text{and}$$

$$\langle\bar{x},\bar{q}\rangle \in c\ell_{X\times\beta Y}\, Z(f);$$

we note that there is a neighborhood V of $\bar{q}$ such that

$$f_{\bar{x}}(y) \geq 1/2 \quad \text{for } y \in V \cap Y$$

and we assume also, replacing f if necessary by the function

$$\langle x,y\rangle \to \min\{1, 2\cdot f(x,y)\},$$

that $\tilde{f}_{\bar{x}}[V \cap Y] \subset \{1\}$.

Now for $n < \omega$ we define $\langle x_n,y_n\rangle \in X \times Y$ and sets U_n, V_n and $\bar{U}_n$ such that

(i) $\langle x_n,y_n\rangle \in Z(f)$ for $n < \omega$;

(ii) $U_n \times V_n$ is a neighborhood of $\langle x_n,y_n\rangle$, and $\bar{U}_n \times V_n$ is a neighborhood of $\langle\bar{x},y_n\rangle$, for $n < \omega$;

(iii) $f(x,y) < 1/3$ for $\langle x,y\rangle \in U_n \times V_n$, $n < \omega$;

(iv) $f(x,y) > 2/3$ for $\langle x,y\rangle \in \bar{U}_n \times V_n$, $n < \omega$; and

(v) $U_{n+1} \cup \bar{U}_{n+1} \subset \bar{U}_n$, for $n < \omega$.

We choose $\langle x_0, y_0 \rangle \in Z(f) \cap (X \times V)$; it is clear how to proceed by recursion.

Since $X \times Y$ is pseudocompact there is $\langle p,q \rangle \in X \times Y$ such that every neighborhood of $\langle p,q \rangle$ intersects infinitely many of the sets $U_n \times V_n$; it follows that $f(p,q) \leq 1/3$. Further if $A \times B$ is a neighborhood of $\langle p,q \rangle$ and $\{n_k : k < \omega\}$ is a sequence of finite cardinals such that

$$(A \times B) \cap (U_{n_k} \times V_{n_k}) \neq \emptyset \quad \text{for} \quad k < \omega$$

then since

$$U_{n_k} \subset \bar{U}_{n_k - 1} \subset \cdots \subset \bar{U}_{n_{k-1}}$$

we have

$$(A \times B) \cap (\bar{U}_{n_k} \times V_{n_k}) \neq \emptyset \quad \text{for} \quad 1 \leq k < \omega;$$

it follows that $f(p,q) \geq 2/3$.

This contradiction shows that $\tilde{f} \in C^*(X \times \beta Y)$. The proof that $X \times Y$ is C^*-embedded in $X \times \beta Y$ is complete.

Since $X \times Y$ is a dense, pseudocompact subspace of $X \times \beta Y$, the space $X \times \beta Y$ is pseudocompact. The argument just given, with βX and βY replacing βY and X respectively, shows that $X \times \beta Y$ is C^*-embedded in $\beta X \times \beta Y$. It follows that $X \times Y$ is C^*-embedded in $\beta X \times \beta Y$, as required.

Remarks. The problem of characterizing those pairs $\langle X,Y \rangle$ of spaces such that $\upsilon(X \times Y) = \upsilon X \times \upsilon Y$ or $\gamma(X \times Y) = \gamma X \times \gamma Y$ is open. Furthermore, with the exception of Theorem 7.6, little seems to be known about the relation $P(X \times Y) = PX \times PY$ for any class P of spaces. Various "exponential laws" of the form

$$C(X \times Y, Z) \subset C(X, C(Y,Z))$$

are available (see for example Noble [69] and the extensive bibliography); we note that to imitate the proof of Lemma 5.2 (with $Z \in P$) it is necessary to topologize $C(Y,Z)$ so that $C(Y,Z) \in P$ and, perhaps, to find a P-analogue of

Lemma 5.1.

Research Problem. Let P be a class of spaces such that $C \subset P = \bar{P}$. Characterize those pairs $\langle X,Y \rangle$ of spaces such that $P(X \times Y) = PX \times PY$.

§8. Absolute (Separable, Metrizable) Borel Spaces

In this and the following Section we characterize those spaces X that are Baire sets in βX and we describe the relation between these spaces and the classical absolute Borel metric spaces of E. Čech. In anticipation of the results and techniques to be used in §9 we introduce more definitions and terminology than are strictly necessary for the present Section.

Definition. Let X be a set. A *σ-algebra* on X is a non-empty family $\mathcal{S}$ of subsets X such that

if $A \in \mathcal{S}$ then $X \backslash A \in \mathcal{S}$, and

if $\{A_n : n < \omega\} \subset \mathcal{S}$ then $\bigcup_{n<\omega} A_n \in \mathcal{S}$ and $\bigcap_{n<\omega} A_n \in \mathcal{S}$.

It is clear that if X is a set then the intersection of σ-algebras on X is a σ-algebra on X, and hence for $\emptyset \neq \mathcal{A} \subset \mathcal{P}(X)$ there is a smallest σ-algebra on X containing $\mathcal{A}$. This justifies (a) below; a similar argument justifies (b).

Notation. Let X be a set and $\emptyset \neq \mathcal{A} \subset \mathcal{P}(X)$.

(a) The smallest σ-algebra on X containing $\mathcal{A}$ is denoted $\sigma\mathcal{A}$ (or sometimes $\sigma_X\mathcal{A}$);

(b) the smallest family $\mathcal{S}$ of subsets of X such that
$\mathcal{A} \subset \mathcal{S}$, and
if $\{A_n : n < \omega\} \subset \mathcal{S}$ then $\bigcup_{n<\omega} A_n \in \mathcal{S}$ and $\bigcap_{n<\omega} A_n \in \mathcal{S}$
is denoted $\tau\mathcal{A}$ (or sometimes $\tau_X\mathcal{A}$).

We recall that if X is a space then $\mathcal{Z}(X)$ denotes the set of zero-sets of X.

Definition. Let X be a space and set

$$\mathcal{C} = \{A \subset X : A \text{ is closed}\} \text{ and}$$
$$\mathcal{Z} = \mathcal{Z}(X).$$

Then

(a) the *Borel sets* of X are the elements of $\sigma_X\mathcal{C}$, and

(b) the *Baire sets* of X are the elements of $\sigma_X\mathcal{Z}$.

For every space X we denote by $\mathcal{B}(X)$ the family of Baire sets of X.

We note that for every space X, every zero-set of X is closed in X; hence every Baire set of X is a Borel set of X. Further, if (X,ρ) is a metric space and A is a closed subset of X then A is the zero-set of the continuous function $x \to \rho(A,x)$; hence every Borel set of a metric space is a Baire set.

For use in the next lemma we set

$$\mathcal{S}' = \{X\backslash A : A \in \mathcal{S}\} \quad \text{for} \quad \mathcal{S} \subset \mathcal{P}(X).$$

8.1. *Lemma.* Let X be a set and let $\mathcal{A} = \mathcal{A}' \subset \mathcal{P}(X)$. Then $\sigma\mathcal{A} = \tau\mathcal{A}$.

Proof. It is enough to prove that $(\tau\mathcal{A})' \subset \tau\mathcal{A}$. Since $\mathcal{A} \subset \tau\mathcal{A}$ we have $\mathcal{A}' \subset (\tau\mathcal{A})'$, and since $(\tau\mathcal{A})'$ is closed under countable union and intersection we have

$$\tau\mathcal{A} = \tau(\mathcal{A}') \subset (\tau\mathcal{A})'$$

and hence $(\tau\mathcal{A})' \subset (\tau\mathcal{A})'' = \tau\mathcal{A}$, as required.

8.2. *Corollary.* Let X be a space and set $\mathcal{Z} = \mathcal{Z}(X)$. Then $\mathcal{B}(X) = \tau\mathcal{Z}$.

Proof. We set $\mathcal{Z}' = \{X\backslash A : A \in \mathcal{Z}\}$ and we note that since

$$\{x \in X : f(x) \neq 0\} = \cup_{n<\omega} \{x \in X : |f(x)| \geq 1/n\}$$

for $f \in C(X)$, we have $Z' \subset \tau Z$ and hence $\tau Z = \tau(Z \cup Z')$. It follows from Lemma 8.1 that

$$\mathcal{B}(X) = \sigma Z = \sigma(Z \cup Z') = \tau(Z \cup Z') = \tau Z.$$

When X is a set and $A \subset P(X)$ it is frequently convenient to define the family τA recursively, instead of as the intersection of certain subsets of $P(X)$. We do this in Lemma 8.3 below. The reader is assumed to be familiar with the following definitions and theorems from the theory of ordinal numbers.

An ordinal number ξ is a *limit* ordinal if there is no ordinal number ζ such that $\xi = \zeta + 1$; in particular, 0 is a limit ordinal.

For every ordinal number ξ there is a largest limit ordinal λ such that $\lambda \le \xi$, and there is a (unique) finite ordinal n such that $\xi = \lambda + n$; the ordinal ξ is said to be *even* if n is even, *odd* if n is odd.

The smallest uncountable cardinal is denoted ω^+.

Notation. Let X be a set and let $A \subset P(X)$. For $\xi < \omega^+$ the set A_ξ is defined by the rule

$$A_0 = A, \text{ and}$$
$$A_\xi = \{\cup_{n<\omega} A_n : A_n \in \cup_{\zeta<\xi} A_\zeta\} \text{ for odd } \xi < \omega^+$$
$$= \{\cap_{n<\omega} A_n : A_n \in \cup_{\zeta<\xi} A_\zeta\} \text{ for even } \xi < \omega^+.$$

8.3. *Lemma.* Let X be a set and $\emptyset \ne A \subset P(X)$. Then

$$\tau A = \cup_{\xi<\omega^+} A_\xi.$$

Proof. It is clear that $\cup_{\xi<\omega^+} A_\xi \subset \tau A$; hence it is enough to prove that $\cup_{\xi<\omega^+} A_\xi$ is closed under countable union and intersection. If

$$\{A_n : n < \omega\} \subset \cup_{\xi<\omega^+} A_\xi$$

then for $n < \omega$ there is $\xi(n) < \omega^+$ such that $A_n \in A_{\xi(n)}$. There is a limit ordinal $\lambda < \omega^+$ such that $\xi(n) < \lambda$ for all $n < \omega$, and we have

$$\cap_{n<\omega} A_n \in A_\lambda \text{ and } \cup_{n<\omega} A_n \in A_{\lambda+1}.$$

8.4. *Corollary.* If X is a set and $\emptyset \neq A \subset P(X)$, then $|\tau A| \leq |A|^\omega$.

Proof. If $|A| = 1$ then $A = \tau A$ and the statement is clearly true. We assume that $|A| \geq 2$ and we claim that

$$|A_\xi| \leq |A|^\omega \text{ for all } \xi < \omega^+.$$

Clearly the claim is true for $\xi = 0$. Further if $\xi < \omega^+$ and if $|A_\zeta| \leq |A|^\omega$ for all $\zeta < \xi$ then from the definition of A_ξ we have

$$|A_\xi| \leq |\cup_{\zeta<\xi} A_\zeta|^\omega \leq (|\xi|\cdot|A|^\omega)^\omega$$
$$\leq (\omega\cdot|A|^\omega)^\omega = |A|^\omega.$$

Thus the claim is proved, and from Lemma 8.3 we have

$$|\tau A| \leq \Sigma_{\xi<\omega^+} |A|^\omega = \omega^+\cdot|A|^\omega = |A|^\omega.$$

8.5. *Corollary.* Let X be a space and set $Z = Z(X)$. Then

(a) $B(X) = \cup_{\xi<\omega^+} Z_\xi$; and

(b) if X is separable then $|B(X)| \leq 2^\omega$; and

(c) if f is a continuous function from X to a space Y and $A \in B(Y)$, then $f^{-1}(A) \in B(X)$.

Proof. (a) follows from Corollary 8.2 and Lemma 8.3.

(b) We note that

$$|Z| \leq |C(X)| \leq |R^\omega| = 2^\omega$$

since X is separable, so that

$$|B(X)| = |\tau Z| \leq |Z|^\omega \leq (2^\omega)^\omega = 2^\omega$$

by Corollary 8.4.

(c) If $A \in Z_0$ there is $g \in C(Y)$ such that $A = Z(g)$, and since $g\circ f \in C(X)$ and

$$f^{-1}(A) = Z(g\circ f)$$

we have $f^{-1}(A) \in Z(X) \subset B(X)$. Now let $\xi < \omega^+$, suppose that

$$f^{-1}(B) \in B(X) \text{ for every } B \in \cup_{\zeta<\xi} Z_\zeta,$$

and let $A \in Z_\xi$. There is $\{A_n : n < \omega\} \subset \cup_{\zeta<\xi} Z_\zeta$ such that either $A = \cup_{n<\omega} A_n$ or $A = \cap_{n<\omega} A_n$, and since $f^{-1}(A_n) \in \mathcal{B}(X)$ for $n < \omega$ and $\mathcal{B}(X)$ is closed under countable unions and countable intersections we have $f^{-1}(A) \in \mathcal{B}(X)$.

It follows from Lemma 8.3 and the axiom of transfinite induction that $f^{-1}(A) \in \mathcal{B}(X)$ for all $A \in \mathcal{B}(Y)$. The proof is complete.

We note from Corollary 8.5(b) that the number of Borel sets of a separable metric space is not greater than 2^ω.

We show by example that the relation "is a Baire subset of" is not transitive. This example suggests the necessity of considering Z-embedded subspaces (as defined below).

We claim first that every countably infinite set has a family of 2^ω infinite subsets of which every two have finite intersection. Let F denote the family of finite subsets of $\omega \times \omega$ (so that $|F| = \omega$) and for $f \in \omega^\omega$ and $n < \omega$ define

$$f|n = \{<k, f(k)> : k < n\} \text{ and}$$

$$S(f) = \{f|n : n < \omega\};$$

then $S(f) \subset F$ for all $f \in \omega^\omega$, and $\{S(f) : f \in \omega^\omega\}$ is a family of 2^ω (infinite) subsets of the countable set F such that

$$|S(f) \cap S(g)| < \omega \text{ if } f, g \in \omega^\omega \text{ and } f \neq g.$$

We recall that for every cardinal number α, the set α with the discrete topology is denoted α.

8.6. *Theorem.* Let $\{A_\xi : \xi < 2^\omega\}$ be a family of infinite subsets of ω such that

$$|A_\xi \cap A_\zeta| < \omega \text{ if } \xi < \zeta < 2^\omega,$$

for $\xi < 2^\omega$ define $\hat{A}_\xi = (c\ell_{\beta(\omega)} A_\xi)\backslash\omega$, and set

$$A = \cup_{\xi<2\omega} \hat{A}_\xi \text{ and } X = \omega \cup A.$$

Then $A \in \mathcal{B}(X)$, and not every Baire set of A is a Baire set of X.

Proof. It is clear that $\beta(\omega)\backslash\omega \in Z(\beta(\omega))$, so that

$$A \in Z(X) \subset B(X).$$

We claim that $\hat{A}_\xi \cap \hat{A}_\zeta = \emptyset$ if $\xi < \zeta < 2^\omega$. Indeed let $f : \omega \to [0,1]$ be defined by the rule

$$f[A_\xi] \subset \{0\},$$
$$f[\omega\backslash A_\xi] \subset \{1\};$$

then $\bar{f}[\hat{A}_\xi] \subset \{0\}$, and $\bar{f}[\hat{A}_\zeta] \subset \{1\}$ since

$$\hat{A}_\zeta = (c\ell_{\beta(\omega)}(A_\zeta\backslash A_\xi))\backslash\omega.$$

It follows from the claim just proved that for every $S \subset 2^\omega$ the sets

$$\cup_{\xi\in S}\, \hat{A}_\xi \quad \text{and} \quad \cup_{\xi\in 2^\omega\backslash S}\, \hat{A}_\xi$$

are disjoint, open subsets of A whose union is A, so that $\cup_{\xi\in S}\, \hat{A}_\xi$ is open-and-closed in A (and hence is an element of $Z(A)$). Thus

$$|B(A)| \geq |Z(A)| \geq |P(2^\omega)| = 2^{2^\omega},$$

and from Corollary 8.5(b) we have $|B(X)| \leq 2^\omega$. The proof is complete.

Definition. Let Y be a space and let $X \subset Y$. Then X is *Z-embedded* in Y if for every $Z \in Z(X)$ there is $Z' \in Z(Y)$ such that $Z = Z' \cap X$.

We note that if X is a C*-embedded subset of a space Y, then X is Z-embedded in Y. Furthermore, since every closed subset of a metric space is a zero-set, we have: Every subset of a metric space is Z-embedded.

8.7. *Lemma.* Let Y be a space and let X be a Z-embedded subspace of Y.

(a) For every $A \in B(X)$ there is $B \in B(Y)$ such that $A = X \cap B$;

(b) if $X \in B(Y)$ then $B(X) \subset B(Y)$.

Proof. (a) We define
$C = \{A \in B(X)\text{: there is } B \in B(Y) \text{ such that } A = X \cap B\}$

and we note that $\mathcal{Z}(X) \subset \mathcal{C} \subset \mathcal{B}(X)$ and hence $\mathcal{B}(X) = \tau\mathcal{C}$. It is clear that if $\{A_n : n < \omega\} \subset \mathcal{C}$ then $\cup_{n<\omega} A_n \in \mathcal{C}$ and $\cap_{n<\omega} A_n \in \mathcal{C}$, so that $\mathcal{C} = \tau\mathcal{C}$. Thus we have $\mathcal{C} = \mathcal{B}(X)$, as required.

(b) follows from (a).

We recall that a compactification of a space X is a compact space in which X is dense. The statement that BX is a metrizable compactification of a (metric) space X does not imply that the inclusion function from X into BX is an isometry.

8.8. *Lemma.* For every space X, the following statements are equivalent.

(a) X has a metrizable compactification;

(b) X is separable and metrizable.

Proof. (a) ⇒ (b). It is well-known and easy to prove that every subspace of a compact metric space is separable and metrizable. We omit the proof.

(b) ⇒ (a). This is obvious if $|X| < \omega$; we assume $|X| \geq \omega$. Let $\langle x_n : n < \omega \rangle$ be a faithfully indexed dense subset of X, for $n < \omega$ let

$$f_n : X \to [0,1]_n = [0,1]$$

be defined by the rule

$$f_n(x) = \min\{1, \rho(x,x_n)\} \quad \text{for } x \in X,$$

set $Y = [0,1]^\omega$ and define $e : X \to Y$ by the rule

$$e(x)_n = f_n(x).$$

It is clear that the function e is one-to-one and continous; thus to verify that e is a homeomorphism onto $e[X]$ it is enough to prove that if $x \in X$ and $\varepsilon > 0$ then $e[S(x,\varepsilon)]$ contains a neighborhood in $e[X]$ of $e(x)$. There is $n < \omega$ such that $\rho(x,x_n) < \varepsilon/2$, and we choose

$$\pi_n^{-1}([0,\varepsilon/2)) \cap e[X]$$

for the required neighborhood.

It is clear that $c\ell_Y\, e[X]$ is a metrizable compactification of $e[X]$; hence X has a metrizable compactification.

8.9. *Lemma.* Let M_0 and M_1 be complete metric spaces and f a homeomorphism from a subspace X of M_0 onto a subspace $f[X]$ of M_1. Then there are G_δ subsets G and H of M_0 and M_1 respectively, and a homeomorphism $\tilde{f}$ of G onto H, such that

$$X \subset G, \quad f[X] \subset H, \quad \text{and} \quad f \subset \tilde{f}.$$

Proof. We assume first that X and $f[X]$ are dense in M_0 and M_1 respectively. By Lemma 3.3 there are G_δ subsets G' and H' of M_0 and M_1 respectively and continuous functions $g : G' \to M_1$ and $h : H' \to M_0$ such that

$$X \subset G', \quad f \subset g, \quad f[X] \subset H', \quad \text{and} \quad f^{-1} \subset h.$$

We define

$$G = G' \cap g^{-1}(H') \quad \text{and} \quad H = H' \cap h^{-1}(G');$$

we note that G is a G_δ subset of G' and hence of M_0, and similarly that H is a G_δ subset of M_1. We set

$$\tilde{g} = g|G \quad \text{and} \quad \tilde{h} = h|H$$

and we claim that $\tilde{h}\circ\tilde{g}$ is a function well-defined on G. We have $\tilde{g}[G] = g[G] \subset H'$, and since $h(g(x)) = x$ for $x \in X$ we have $h(g(p)) = p$ for $p \in G$ by Lemma 1.4 and hence

$$\tilde{g}[G] = g[G] \subset h^{-1}(G) \subset h^{-1}(G');$$

the claim is proved. It follows that $\tilde{h}\circ\tilde{g} = id_G$ and (by a similar proof) that $\tilde{g}\circ\tilde{h} = id_H$. Hence $\tilde{h} = \tilde{g}^{-1}$ and $\tilde{g}$ is a homeomorphism. We define $\tilde{f} = \tilde{g}$.

For the general case we note that since X and $f[X]$ are dense in the (complete metric) spaces $c\ell_{M_0} X$ and $c\ell_{M_1} f[X]$ respectively, there are G_δ subsets G and H of $c\ell_{M_0} X$ and $c\ell_{M_1} f[X]$ and a homeomorphism $\tilde{f}$ of G onto H such that $f \subset \tilde{f}$. Since $c\ell_{M_0} X$ and $c\ell_{M_1} f[X]$ are G_δ subsets of M_0 and M_1, the sets G and H are G_δ subsets of M_0 and M_1. The proof is complete.

Definition. Let X be a separable metric space. Then X is an *absolute Borel space* if $X \in \mathcal{B}(BX)$ for all metrizable compactifications BX of X.

8.10. *Theorem.* Every completely metrizable, separable space is an absolute Borel space.

Proof. Let BX be a metrizable compactification of the completely metrizable, separable space X. Then X is a G_δ in BX by Theorem 3.6 and Lemma 3.5; since BX is metrizable we have $X \in \mathcal{B}(BX)$ as required.

8.11. *Theorem.* If X is a separable metric space, the following statements are equivalent.

(a) X is an absolute Borel space;

(b) if K is a compact metric space and $X \subset K$, then $X \in \mathcal{B}(K)$;

(c) there is a metrizable compactification BX of X such that $X \in \mathcal{B}(BX)$.

Proof. (a) $\Rightarrow$ (b). We set $BX = c\ell_K X$. Since $X \in \mathcal{B}(BX)$ by (a) we have $X \in \mathcal{B}(K)$ by Lemma 8.7(b).

(b) $\Rightarrow$ (c) follows from Lemma 8.8((b) $\Rightarrow$ (a)).

(c) $\Rightarrow$ (a). Let BX and K be metrizable compactifications of X, and let $X \in \mathcal{B}(BX)$. By Lemma 8.9 there are G_δ subsets G and H of K and BX respectively and a homeomorphism f of G onto H such that

$X \subset G$, $X \subset H$, and

$f(x) = x$ for all $x \in X$.

Since $X \in \mathcal{B}(BX)$ we have $X \in \mathcal{B}(H)$ and hence $X \in \mathcal{B}(G)$; then since $G \in \mathcal{B}(K)$ and G is Z-embedded in K we have $X \in \mathcal{B}(K)$ (from Lemma 8.7(b)), as required.

8.12. *Corollary.* Let $\{X_n : n < \omega\}$ be a countable family of (separable metric) absolute Borel spaces, and set

$X = \prod_{n<\omega} X_n$. Then X is an absolute Borel space.

Proof. By Theorem 8.11, for $n < \omega$ there is a metrizable compactification BX_n of X_n such that $X_n \in \mathcal{B}(BX_n)$. We define

$$BX = \prod_{n<\omega} BX_n,$$

we denote by π_n the projection function from BX onto BX_n, and we note that since $X_n \in \mathcal{B}(BX_n)$ we have $\pi_n^{-1}(X_n) \in \mathcal{B}(BX)$ by Corollary 8.5(c) and hence

$$X = \bigcap_{n<\omega} \pi_n^{-1}(X_n) \in \mathcal{B}(BX).$$

Since BX is a metrizable compactification of X, the result now follows from Theorem 8.11.

§9. TOPOLOGICAL PROPERTIES OF BAIRE SETS

We here complete the results established in §8, describing (*via* perfect functions) the relation between those spaces X that are Baire in βX and the absolute Borel spaces.

The two principal results of this Section are these: Every Baire set in a compact space is a Lindelöf space (Theorem 9.10(a)), and every space that is Baire in one of its compactifications is Baire in each of its compactifications (Theorems 9.6 and 9.12).

We prove in Theorem 9.14 that several of the classes of spaces we consider are closed under the formation of countable products, and in §§9.15 - 9.17 we prove that every compact Baire set, and indeed in a surprisingly large class of spaces every closed Baire set, is a zero-set.

9.1. *Lemma.* Let Y be a space and $A, B \in \mathcal{B}(Y)$. Then there are $\mathcal{W} \subset \mathcal{Z}(Y)$, a metric space K and a continuous funtion g from Y onto K such that

$|\mathcal{W}| \leq \omega$,

$A, B \in \tau\mathcal{W}$, and

if $C \in \tau\mathcal{W}$ then $g[C] \in \mathcal{B}(K)$ and $C = g^{-1}(g[C])$.

Proof. Let $\mathcal{Y} = \cup\{\tau\mathcal{W} : \mathcal{W} \subset \mathcal{Z}(Y), |\mathcal{W}| \leq \omega\}$. It is clear that $\mathcal{Z}(Y) \subset \mathcal{Y} \subset \mathcal{B}(Y)$ and $\tau\mathcal{Y} = \mathcal{Y}$, so that $\mathcal{Y} = \mathcal{B}(Y)$. Hence there is $\mathcal{W} \subset \mathcal{Z}(Y)$ such that $|\mathcal{W}| \leq \omega$ and $A, B \in \tau\mathcal{W}$. Let

$<Z_n : n < \omega>$

be an enumeration of $\mathcal{W}$, for $n < \omega$ let f_n be a continuous function from Y into $[0,1]$ such that $Z_n = Z(f_n)$, define

$g : Y \rightarrow [0,1]^\omega$

by the rule

$$g(y)_n = f_n(y) \quad \text{for all} \quad y \in Y,$$

and set $K = g[Y]$. Then g is continuous from Y onto K, and for $x, y \in Y$ we have

$$g(x) = g(y) \quad \text{if and only if} \quad f_n(x) = f_n(y) \quad \text{for} \quad n < \omega.$$

We recall from Lemma 8.3 that $\tau\mathcal{W} = \cup_{\xi<\omega^+} \mathcal{W}_\xi$. Thus to complete the proof it is enough to prove that if $\xi < \omega^+$ and $C \in \mathcal{W}_\xi$, then $g[C] \in \mathcal{B}(K)$ and $C = g^{-1}(g[C])$. We proceed by transfinite recursion.

Let $C \in \mathcal{W}_0$; there is $k < \omega$ such that $C = Z_k$. If $y \in Y$ and $g(y) \notin g[C]$, then $f_k(y) \neq 0$ and $\{p \in K : p_k \neq 0\}$ is a neighborhood of $g(y)$ in K disjoint from $g[C]$; it follows that $g[C]$ is closed in K, so that $g[C] \in \mathcal{B}(K)$. Further for every $y \in Y \backslash C$ we have $f_k(y) \neq 0$ and hence $y \notin g^{-1}(g[C])$; it follows that $C = g^{-1}(g[C])$.

Now let $\xi < \omega^+$ and $C \in \mathcal{W}_\xi$, and assume that if $D \in \cup_{\zeta<\xi} \mathcal{W}_\zeta$ then $g[D] \in \mathcal{B}(K)$ and $D = g^{-1}(g[D])$. By the definition of $\mathcal{W}_\xi$ there is

$$\{C_n : n < \omega\} \subset \cup_{\zeta<\xi} \mathcal{W}_\zeta$$

such that $C = \cup_{n<\omega} C_n$ or $C = \cap_{n<\omega} C_n$. Since

$$C_n = g^{-1}(g[C_n]) \quad \text{for} \quad n < \omega$$

we have

$$\begin{aligned} g[\cap_{n<\omega} C_n] &= g[\cap_{n<\omega} g^{-1}(g[C_n])] \\ &= g[g^{-1}(\cap_{n<\omega} g[C_n])] = \cap_{n<\omega} g[C_n] \end{aligned}$$

and hence

$$\begin{aligned} g^{-1}(g[\cap_{n<\omega} C_n]) &= g^{-1}(\cap_{n<\omega} g[C_n]) \\ &= \cap_{n<\omega} g^{-1}(g[C_n]) = \cap_{n<\omega} C_n; \end{aligned}$$

and similarly

$$g[\cup_{n<\omega} C_n] = \cup_{n<\omega} g[C_n] \quad \text{and}$$

$$g^{-1}(g[\cup_{n<\omega} C_n]) = \cup_{n<\omega} C_n.$$

It follows that $C = g^{-1}(g[C])$ and (since $g[C_n] \in \mathcal{B}(K)$ for $n < \omega$) that $g[C] \in \mathcal{B}(K)$.

PERFECT FUNCTIONS

We recall that a function $f : X \to Y$ is said to be a *closed* function if $f[A]$ is closed in Y for every closed subset A of X.

Definition. Let X and Y be spaces and f a function from X onto Y. Then f is a *perfect* function if f is a continuous, closed function such that $f^{-1}(\{y\})$ is compact for every $y \in Y$.

9.2. *Lemma.* Let X and Y be spaces and f a continuous function from X onto Y. The following statements are equivalent.

(a) f is a perfect function;

(b) $\bar{f}[\beta X \backslash X] = BY \backslash Y$ for every compactification BY of Y;

(c) $\bar{f}[\beta X \backslash X] = \beta Y \backslash Y$;

(d) there are compactifications BX and BY of X and Y respectively and a continuous function $\tilde{f} : BX \to BY$ such that $f \subset \tilde{f}$ and $\tilde{f}[BX \backslash X] = BY \backslash Y$.

Proof. (a) $\Rightarrow$ (b). Since $\bar{f}[\beta X]$ is compact and dense in BY, we have $\bar{f}[\beta X \backslash X] \supset BY \backslash Y$. If there is $p \in \beta X \backslash X$ such that $\bar{f}(p) \in Y$ then since $f^{-1}(\{\bar{f}(p)\})$ is a compact subset of X there is an open subset U of X such that

$$f^{-1}(\{\bar{f}(p)\}) \subset U \text{ and } p \notin c\ell_{\beta X}\, U.$$

Then $p \in c\ell_{\beta X}(X \backslash U)$, and hence

$$\bar{f}(p) \in c\ell_{BY}\, \bar{f}[X \backslash U] = c\ell_{BY}\, f[X \backslash U];$$

since $\bar{f}(p) \in Y$ and $f[X \backslash U] \subset Y$ we have

$$\bar{f}(p) \in c\ell_{Y}\, f[X \backslash U] = f[X \backslash U],$$

contradicting the inclusion $f^{-1}(\{\bar{f}(p)\}) \subset U$.

(b) $\Rightarrow$ (c) is obvious.

(c) $\Rightarrow$ (d) is obvious.

(d) ⇒ (a). For every closed subset A of X we have

$$f[A] = \tilde{f}[c\ell_{\beta X} A] \cap Y,$$

and hence f is a closed function (from X onto Y). Further for $y \in Y$ we have

$$f^{-1}(\{y\}) = \bar{f}^{-1}(\{y\}) \subset X,$$

so that $f^{-1}(\{y\})$ is compact.

9.3. *Corollary.* Let X and Y be spaces, and f a perfect function from X onto Y.

(a) If Y is compact, then X is compact;

(b) if Y is a Lindelöf space, then X is a Lindelöf space;

(c) if Y is realcompact, then X is realcompact;

(d) if Y is paracompact, then X is paracompact;

(e) if Y is topologically complete, then X is topologically complete;

(f) if Y is σ-compact, then X is σ-compact; and

(g) if Y is a G_δ in βY, then X is a G_δ in βX.

Proof. Let $\bar{f}$ denote the continuous function from βX onto βY such that $f \subset \bar{f}$. Since $\bar{f}[\beta X \backslash X] = \beta Y \backslash Y$ by Lemma 9.2, we have $X = \bar{f}^{-1}(Y)$ and hence (by Lemma 1.9) X is homeomorphic to a closed subspace of $\beta X \times Y$. Statements (a) and (f) are now clear, and (b), (c), (d) and (e) follow from Corollary 4.5. Finally to prove (g) we note that if $\{U_n : n < \omega\}$ is a family of open subsets of βY such that $Y = \cap_{n<\omega} U_n$, then $\bar{f}^{-1}(U_n)$ is an open subset of βX for all $n < \omega$ and

$$X = \bar{f}^{-1}(Y) = \bar{f}^{-1}(\cap_{n<\omega} U_n) = \cap_{n<\omega} \bar{f}^{-1}(U_n).$$

The proof is complete.

We recall from Lemma 3.5 that a space X is a G_δ in βX if and only if X is a G_δ in every compactification of X.

9.4. *Theorem.* For every space X, the following statements are equivalent.

(a) X is paracompact, and a G_δ in βX;

(b) X is paracompact and there is a compact space K such that X is a G_δ in K;

(c) there is a perfect function from X onto a complete metric space.

Proof. (a) ⇒ (b) is obvious.

(b) ⇒ (a). Clearly X is a G_δ in $c\ell_K X$. Hence X is a G_δ in βX by Lemma 3.5.

(a) ⇒ (c). There is a family $\{K_n : n < \omega\}$ of compact subsets of βX such that $\beta X\backslash X = \cup_{n<\omega} K_n$, and by Theorem 4.4(c) there is for $n < \omega$ a locally finite cozero cover $\mathcal{U}_n$ of X such that $U \in \mathcal{U}_n$ implies $K_n \cap c\ell_{\beta X} U = \emptyset$. We set $\mathcal{U} = \cup_{n<\omega} \mathcal{U}_n$ and we define a metric space (M,ρ) and a continuous function f from X onto M as in the statement of Theorem 3.2. We note that if $p \in \beta X\backslash X$ then there is a cover $\mathcal{V}(p) \subset \mathcal{U}$ such that $p \notin \cup\{c\ell_{\beta X} U : U \in \mathcal{V}(p)\}$; indeed there is $n < \omega$ such that $p \in K_n$ and $\mathcal{U}_n$ is such a cover. We denote by $\bar{f}$ the continuous function from βX onto βM such that $f \subset \bar{f}$; it follows from Theorem 3.2(d) that $\bar{f}[\beta X\backslash X] = \beta M\backslash M$, so that f is a perfect function by Lemma 9.2. Further

$$\beta M\backslash M = \cup_{n<\omega} \bar{f}[K_n],$$

so that $\beta M\backslash M$ is σ-compact and hence M is a G_δ in βM. Hence M is completely metrizable by Theorem 3.6.

(c) ⇒ (a). A complete metric space M is paracompact, and a G_δ in βM, by Theorems 3.1 and 3.6. Thus X is paracompact, and a G_δ in βX, by Corollary 9.3(d) and (g).

9.5. *Corollary.* For every space X, the following statements are equivalent.

(a) X is a Lindelöf space, and a G_δ in βX;

(b) X is a Lindelöf space and there is a compact space K such that X is a G_δ in K;

(c) there is a perfect function from X onto a complete, separable metric space.

Proof. (a) ⇒ (b) is obvious.

(b) ⇒ (c). Since X is paracompact by Lemma 4.2(a), there is by Theorem 9.4 a perfect function f from the Lindelöf space X onto a complete metric space M; M is a Lindelöf metric space, hence separable.

(c) ⇒ (a). A complete, separable metric space M is a Lindelöf space, and (by Theorem 3.6) a G_δ in βM. Hence X is Lindelöf, and a G_δ in βX, by Corollary 9.3(b) and (g).

9.6. *Theorem.* For every space X, the following statements are equivalent.

(a) X is a Baire set in βX, *i.e.*, $X \in \mathcal{B}(\beta X)$;

(b) there is a compact space Y such that $X \in \mathcal{B}(Y)$;

(c) there are a separable metric, absolute Borel space M and a perfect function from X onto M.

Proof. (a) ⇒ (b) is obvious.

(b) ⇒ (c). From Lemma 9.1 (with A and B replaced by X) there are a metric space K and a continuous function g from Y onto K such that $X = g^{-1}(g[X])$ and $g[X] \in \mathcal{B}(K)$. We define

$$BX = c\ell_Y X, \quad M = g[X], \quad \text{and} \quad BM = c\ell_K M,$$
$$f = g|X \quad \text{and} \quad \tilde{f} = g|BX.$$

We note that

BX and BM are compactifications of X and M

respectively,

f is a continuous function from X onto M, and

$\tilde{f}$ is a continuous function from BX onto BM such that $f \subset \tilde{f}$ and $\tilde{f}[BX \backslash X] = BM \backslash M$.

It follows from Lemma 9.2((d) ⇒ (a)) that f is a perfect function (from X onto M). Finally since $M \subset BM \subset K$ and $M \in \mathcal{B}K$ we have $M \in \mathcal{B}(BM)$, so that M is a separable metric, absolute Borel space by Theorem 8.11((c) ⇒ (a)).

(c) ⇒ (a). If f is a perfect function from X onto M then from Lemma 9.2((a) ⇒ (c)) we have $\bar{f}[\beta X \backslash X] = \beta M \backslash M$ and hence

$$X = \bar{f}^{-1}(f[X]) = \bar{f}^{-1}(M) \in \mathcal{B}(\beta X)$$

from Corollary 8.5(c).

INHERITED PROPERTIES OF BAIRE SETS

In Theorem 9.10 below we give two proofs that every Baire set in a compact space is a Lindelöf space. The first of these is essentially an appeal to Theorem 9.6. We now give the lemma required for the second.

9.7. *Lemma.* Let Y be a space and $X \subset Y$ and suppose there is a countable family $\mathcal{A}$ of compact subsets of Y such that $X \subset \cup\mathcal{A}$ and

$$\cap\{A \in \mathcal{A} : x \in A\} \subset X \quad \text{for every } x \in X.$$

Then X is a Lindelöf space.

Proof. We suppose without loss of generality that if $\mathcal{F} \subset \mathcal{A}$ and $|\mathcal{F}| < \omega$ then $\cap\mathcal{F} \in \mathcal{A}$. Let $\mathcal{U}$ be a cover of X by (relatively) open subsets of X and let $\mathcal{V}$ be a family of open subsets of Y such that $\mathcal{U} = \{V \cap X : V \in \mathcal{V}\}$. For $x \in X$ the family

$$\{A \in \mathcal{A} : x \in A\} \cup \{Y \backslash \cup\mathcal{V}\}$$

does not have the finite intersection property and hence for every $x \in X$ there is $A_x \in \mathcal{A}$ such that

$$x \in A_x \subset \cup\mathcal{V}.$$

Since $\{A_x : x \in X\}$ is a countable family and every A_x is compact, there is $\{V_n : n < \omega\} \subset \mathcal{V}$ such that

$$\cup_{x \in X} A_x \subset \cup_{n<\omega} V_n;$$

it is clear that $\{V_n \cap X : n < \omega\}$ is a countable cover of X (by elements of $\mathcal{U}$).

9.8. *Corollary.* Let Y be compact and $\mathcal{C}$ the family of closed subsets of Y. If $X \in \tau\mathcal{C}$ then X is a Lindelöf space.

Proof. We have $\tau\mathcal{C} = \cup_{\xi<\omega^+} \mathcal{C}_\xi$ by Lemma 8.3, so by Lemma 9.6 it is enough to prove the following statement:

(*) if $\xi < \omega^+$ and $X \in \mathcal{C}_\xi$, there is $\mathcal{A} \subset \mathcal{C}$ such that $|\mathcal{A}| \le \omega$, $X \subset \cup\mathcal{A}$, and

$$\cap\{A \in \mathcal{A} : x \in A\} \subset X \quad \text{for every } x \in X.$$

If $\xi = 0$ statement (*) is proved by setting $\mathcal{A} = \{X\}$. Now let $\zeta < \omega^+$, assume that (*) holds for $\xi < \zeta$, and let $X \in \mathcal{C}_\zeta$. There is $\{X_n : n < \omega\} \subset \cup_{\xi<\zeta} \mathcal{C}_\xi$ such that

$$X = \cup_{n<\omega} X_n \quad \text{or} \quad X = \cap_{n<\omega} X_n,$$

and for $n < \omega$ there is $\mathcal{A}_n \subset \mathcal{C}$ as in (*); we define $\mathcal{A} = \cup_{n<\omega} \mathcal{A}_n$ and we have, using the properties of the families $\mathcal{A}_n$, that $\mathcal{A}$ is a family as in (*) for X. The proof is complete.

The following lemma is needed for the proof of Theorem 9.10(f).

9.9. *Lemma.* If Y is a locally compact space, the following statements are equivalent.

(a) Y is paracompact;

(b) there is a family $\{Y_i : i \in I\}$ of pairwise disjoint, open-and-closed, σ-compact subsets of Y such that $Y = \cup_{i \in I} Y_i$.

Proof. (a) ⇒ (b). There is a locally finite open cover $\mathcal{U}$ of Y such that $cl_Y U$ is compact for every $U \in \mathcal{U}$. We define a relation $\sim$ on $\mathcal{U}$ as follows: If $U, U' \in \mathcal{U}$ then $U \sim U'$ if there are $n < \omega$ and $\{U_0, U_1, \ldots, U_n\} \subset \mathcal{U}$ such that

$$U = U_0, \quad U' = U_n, \quad \text{and} \quad U_k \cap U_{k+1} \neq \emptyset \quad \text{for} \quad k < n.$$

It is clear that $\sim$ is an equivalence relation on $\mathcal{U}$. We let $\{\mathcal{U}(i) : i \in I\}$ be an indexing of the equivalence classes and for $i \in I$ we define

$$Y_i = \cup\, \mathcal{U}(i).$$

It is clear that the sets Y_i are pairwise disjoint; further, since $\mathcal{U}(i) \subset \mathcal{U}$, every set Y_i is open and hence closed. If $i \in I$ and $U \in \mathcal{U}(i)$ then for every $x \in cl_Y U$ there is a neighborhood V_x of x such that

$$|\{U' \in \mathcal{U} : V_x \cap U' \neq \emptyset\}| < \omega;$$

since $cl_Y U$ is compact it follows that

$$|\{U' \in \mathcal{U} : U \cap U' \neq \emptyset\}| < \omega,$$

and hence $|\mathcal{U}(i)| \leq \omega$. Then since

$$Y_i = \cup\, \mathcal{U}(i) = \cup\{cl_Y U : U \in \mathcal{U}(i)\} \quad \text{for} \quad i \in I,$$

every set Y_i is σ-compact.

(b) ⇒ (a). The σ-compact sets Y_i are Lindelöf sets and hence paracompact (by Lemma 4.2(a)). It is clear that their "disjoint union" Y is paracompact.

9.10. *Theorem*. Let Y be a space and $X \in \mathcal{B}(Y)$.

(a) If Y is compact, then X is a Lindelöf space;

(b) if Y is realcompact, then X is realcompact;

(c) if Y is topologically complete, then X is topologically complete;

(d) if Y is σ-compact, then X is a Lindelöf space;

(e) if Y is a locally compact Lindelöf space, then X is a Lindelöf space;

(f) if Y is locally compact and paracompact, then X

is paracompact;

(g) if $Y \in \mathcal{B}(\beta Y)$, then X is a Lindelöf space.

Proof. (a) (First Proof). By Theorem 9.6((b) ⇒ (c)) there is a perfect function from X onto a separable metric (Lindelöf) absolute Borel space; thus X is a Lindelöf space by Corollary 9.3(b).

(a) (Second Proof). Let $\mathcal{C}$ denote the family of closed subsets of Y. Since

$$X \in \mathcal{B}(Y) = \tau(\mathcal{Z}(Y)) \subset \tau\mathcal{C},$$

X is a Lindelöf space by Corollary 9.8.

(b) Y is $\mathcal{Z}$-embedded in βY and hence by Lemma 8.7(a) there is $X' \in \mathcal{B}(\beta Y)$ such that $X = X' \cap Y$. The space X' is Lindelöf by (a), and hence realcompact by Corollary 2.4. It follows from Lemma 1.8 that X is realcompact.

The proof of (c) is similar to the proof of (b).

(d) follows from (a).

(e) is a special case of (d).

(f) follows from (e) and Lemma 9.9.

(g) Since $X \in \mathcal{B}(\beta Y)$ by Lemma 8.7(b), this follows from (a).

Remark. It is natural to ask whether a Baire set in a (not necessarily locally compact) Lindelöf space must be a Lindelöf space, and whether a Baire set in a paracompact space must be paracompact. We discuss these questions in §11 below.

We have seen in Theorem 9.6 that if a space X is a Baire set in one of its compactifications then X is a Baire set in βX. We now show that such a space is Baire in each of its compactifications; this result (Theorem 9.12) is, together with Theorem 9.6, the correct analogue (for arbitrary spaces) of Theorem 8.11 (for separable metric spaces).

9.11. *Lemma.* Let Y be a space and X a Lindelöf subspace of Y. Then

(a) X is Z-embedded in Y; and

(b) if G is an open subset of Y such that $X \subset G$, then there is a cozero subset U of Y such that $X \subset U \subset G$.

Proof. (a) It is enough to prove that for every cozero set U of X there is a cozero set V of Y such that $U = V \cap X$. For every $x \in U$ there is a cozero set V_x of Y such that

$$x \in V_x \cap X \subset U;$$

since U is the union of a countable family of closed subsets of X, U is a Lindelöf space and hence there is $\{x_n : n < \omega\} \subset U$ such that

$$U = \cup_{n<\omega}(V_{x_n} \cap X) = (\cup_{n<\omega} V_{x_n}) \cap X.$$

We define $V = \cup_{n<\omega} V_{x_n}$.

(b) For every $x \in X$ there is a cozero set U_x of Y such that

$$x \in U_x \subset G,$$

and since X is a Lindelöf space there is $\{x_n : n < \omega\} \subset X$ such that $X \subset \cup_{n<\omega} U_{x_n}$. We define $U = \cup_{n<\omega} U_{x_n}$.

We remark that it follows from Theorem 9.10(a) and Lemma 9.11(a) that every Baire set of a compact space is Z-embedded.

Definition. Let X be a space. Then X is an *absolute Baire space* if $X \in \mathcal{B}(BX)$ for every compactification BX of X.

9.12. *Theorem.* For every space X, the following statements are equivalent.

(a) X is an absolute Baire space;

(b) X is a Baire set in βX;

(c) X is a Baire set in some compactification of X.

Proof. (a) $\Rightarrow$ (b) is obvious.

The equivalence (b) $\Leftrightarrow$ (c) is given by Theorem 9.6.

(b) $\Rightarrow$ (a). Let BX be a compactification of X; we prove that $X \in \mathcal{B}(BX)$. By Lemma 9.1 there are a (compact) metric space K and a continuous function g from βX onto K such that $g[X] \in \mathcal{B}(K)$ and $X = g^{-1}(g[X])$. We define

$$M = g[X] \quad \text{and} \quad f = g|X,$$

so that f is a perfect function from X onto M (by Lemma 9.2) and M is an absolute Borel space (by Theorem 8.11((b) $\Rightarrow$ (a))).

Since $f[X] = M \subset K$ and K is a complete metric space there are by Lemma 3.3 a G_δ set G in BX and a continuous function $\tilde{f} : G \to K$ such that

$$X \subset G \quad \text{and} \quad f \subset \tilde{f}.$$

We claim first that X is a Baire set in G. To verify this it is by Corollary 8.5(c) enough to prove that $X = \tilde{f}^{-1}(M)$. Let h be the (unique) continuous function from βX onto BX such that

$$h(x) = x \quad \text{for} \quad x \in X,$$

and set $H = h^{-1}(G)$. Then $\tilde{f}\circ(h|H)$ and $g|H$ are well-defined continuous functions from H into K, and for $x \in X$ we have

$$(\tilde{f}\circ h)(x) = f(x) \quad \text{and} \quad g(x) = f(x);$$

since X is dense in H it follows from Lemma 1.4 that $(\tilde{f}\circ h)(q) = g(q)$ for $q \in H$. Suppose now that $X \neq \tilde{f}^{-1}(M)$. Then there are $p \in G\backslash X$ and $x \in X$ such that $\tilde{f}(p) = f(x)$, and since $h[H] = G$ there is $q \in H$ such that $h(q) = p$. It follows that

$$g(q) = (\tilde{f}\circ h)(q) = \tilde{f}(p) = f(x) \in M,$$

so that $q \in X$ (since $X = g^{-1}(g[X])$) and hence

$$p = h(q) = q \in X,$$

a contradiction. The proof that X is a Baire set in G

is complete.

We note from (b) and Theorem 9.10(a) that X is a Lindelöf space, so from Lemma 9.11(b) there is a countable family $\{U_n : n < \omega\}$ of cozero sets of BX such that

$$X \subset \bigcap_{n<\omega} U_n \subset G.$$

We define $U = \bigcap_{n<\omega} U_n$. Then $U \in \mathcal{B}(BX)$, so that U is a Lindelöf space by Theorem 9.10(a) and hence $\mathcal{Z}$-embedded in BX by Lemma 9.11(a). Since $X \subset U \subset G$ and $X \in \mathcal{B}(G)$ we have $X \in \mathcal{B}(U)$, so that $X \in \mathcal{B}(BX)$ by Lemma 9.7(b). The proof is complete.

SOME PRODUCT-SPACE THEOREMS

9.13. ***Lemma.*** Let I be a set and for $i \in I$ let X_i and Y_i be spaces and f_i a perfect function from X_i onto Y_i. Set $X = \prod_{i\in I} X_i$ and $Y = \prod_{i\in I} Y_i$, and define $f : X \to Y$ by the rule

$$f(x)_i = f_i(x_i) \quad \text{for} \quad x \in X.$$

Then f is a perfect function from X onto Y.

Proof. It is clear that f is a well-defined function from X onto Y.

Set $BX = \prod_{i\in I} \beta X_i$ and $BY = \prod_{i\in I} \beta Y_i$, for $i \in I$ let $\overline{f_i}$ be the continuous function from βX_i onto βY_i such that $f_i \subset \overline{f_i}$, and define $\tilde{f} : BX \to BY$ by

$$\tilde{f}(p)_i = \overline{f_i}(p_i) \quad \text{for} \quad p \in BX.$$

Let π_i and η_i denote respectively the projection functions from BX onto X_i and BY onto Y_i; then

$$\eta_i \circ \tilde{f} = \overline{f_i} \circ \pi_i \quad \text{for} \quad i \in I$$

and hence $\tilde{f}$ is continuous. It follows that $f[BX] = BY$, and hence $BY\backslash Y \subset \tilde{f}[BX\backslash X]$. Further if $p \in BX\backslash X$ there is $i \in I$ such that $p_i \in \beta X_i\backslash X_i$ and hence

$$\tilde{f}(p)_i = \overline{f_i}(p_i) \in \beta Y_i\backslash Y_i$$

by Lemma 9.2; thus $\tilde{f}(p) \in BY\backslash Y$. We have $f[BX\backslash X] = BY\backslash Y$, and hence f is a perfect function by Lemma 9.2.

9.14. *Theorem.* Let $\{X_n : n < \omega\}$ be a countable family of spaces and set $X = \prod_{n<\omega} X_n$.

(a) If X_n is a G_δ in βX_n for $n < \omega$, then X is a G_δ in βX;

(b) if X_n is σ-compact for $n < \omega$, then X is a Lindelöf space;

(c) if X_n is paracompact and a G_δ in βX_n for $n < \omega$, then X is paracompact and a G_δ in βX;

(d) if X_n is a Lindelöf space and a G_δ in βX_n for $n < \omega$, then X is a Lindelöf space and a G_δ in βX;

(e) if X_n is a Baire set in βX_n for $n < \omega$, then X is a Baire set in βX.

Proof. To prove (a) and (b) we set $BX = \prod_{n<\omega} \beta X_n$, we denote by π_n the projection function from BX onto βX_n, and we note that $X = \bigcap_{n<\omega} \pi_n^{-1}(X_n)$. In (a) the set $\pi_n^{-1}(X_n)$ is a G_δ in BX for $n < \omega$, so X is a G_δ in BX and statement (a) follows from Lemma 3.5; in (b) the set $\pi_n^{-1}(X_n)$ is a σ-compact subspace of BX for $n < \omega$, so statement (d) follows from Corollary 9.8.

In (c), (d) and (e) there are for $n < \omega$ (by 9.4, 9.5 and 9.6 respectively) a metric space M_n and a perfect function f_n from X_n onto M_n; further M_n may be chosen complete in (c), complete and separable in (d), and (separable) absolute Borel in (e). We define $M = \prod_{n<\omega} M_n$ and $f : X \to M$ by the rule

$$f(x)_n = f_n(x_n) \quad \text{for} \quad x \in X$$

and we note from Lemma 9.13 that f is a perfect function from X onto M. Since M is complete in (c), complete and separable in (d), and (separable) absolute Borel in (e), statements (c), (d) and (e) now follow from Theorem 9.4, Corollary 9.5, and Theorem 9.6 respectively.

CLOSED BAIRE SETS

According to Corollary 8.5(a), every Baire set in a space X is, for some $\xi < \omega^+$, an element of the class Z_ξ (where $Z = Z(X)$). One may consider when it occurs that an element of one of the classes Z_ξ is a closed subset of X. The rather surprising answer is that for many spaces X every closed Baire set of X is an element of Z_0, *i.e.*, is a zero-set of X.

9.15. *Theorem.* Let X be a Baire set of βX and let A be a closed Baire set of X. Then $A \in Z(X)$.

Proof. We have $A \in B(\beta X)$ by Lemma 8.7(b), so by Lemma 9.1 (with B and Y replaced by X and βX, respectively) there are a (compact) metric space K and a continuous function g from βX onto K such that

$$A = g^{-1}(g[A]) \quad \text{and} \quad X = g^{-1}(g[X]).$$

We define $f = g|X$. Then $\bar{f} = g$ and hence f is a perfect function by Lemma 9.2. It follows that $f[A]$ is a closed subset of the metric space $f[X]$, so that $f[A] \in Z(f[X])$. Since $A = f^{-1}(f[A])$ we have $A \in Z(X)$, as required.

9.16. *Corollary.* If X satisfies one of the following conditions, then every closed Baire of X is a zero-set of X.

(a) X is locally compact and σ-compact;

(b) X is a Lindelöf space, and a G_δ in βX;

(c) X is locally compact and paracompact.

Proof. In (a) the space X is clearly equal to the union of a countable family of cozero sets of βX, and in (b) the space X is (by Lemma 9.11(b)) equal to the intersection of a countable family of cozero sets of βX. Thus in (a) and (b) we have $X \in B(\beta X)$ and the result follows from Theorem 9.15.

(c) follows from (a) and Lemma 9.9.

We remark that it is not known whether every closed Baire set of a locally compact normal space is a zero-set. We show by an example in Theorem 11.1 below that not every closed Baire set is a zero-set.

We conclude this Section with a result similar to Theorem 9.15.

9.17. *Theorem.* If X is a space and A is a pseudocompact Baire set of X, then $A \in Z(X)$.

Proof. By Lemma 9.1 there are a metric space K and a continuous function g from X onto K such that $A = g^{-1}(g[A])$. Since A is pseudocompact and g is continuous the space $g[A]$ is a pseudocompact metric space; it follows from the remarks after Lemma 7.5 that $g[A]$ is compact. Then $g[A] \in Z(K)$, and since $A = g^{-1}(g[A])$ we have $A \in Z(X)$, as required.

§10. LOCAL CONNECTEDNESS IN βX

We prove in this Section that X is locally connected if and only if γX is locally connected, and that βX is locally connected if and only if X is both locally connected and pseudocompact. We begin with an informal listing of several familiar definitions and theorems.

A space is *connected* if it is not equal to the union of two non-empty, disjoint open subsets. If $x \in X$ and $\{K_i : i \in I\}$ is a family of connected subsets of X each containing x, then $\cup_{i \in I} K_i$ is connected. A *component* of X is a maximal, connected subspace of X. Since $c\ell_X K$ is connected if K is a connected subspace of X, every component of a space X is closed. A space X is *locally connected at* x if every neighborhood in X of x contains an open, connected neighborhood of x. A space locally connected at each of its points is *locally connected.* Thus a space is locally connected if and only if every component of every open set is open. (Indeed if U is an open neighborhood in the locally connected space X of x, and if y is an element of the component K of U containing x, then some connected, open subset V of U contains y and hence $K \cup V$, being a connected space between K and U, coincides with K; conversely if U is an open neighborhood of x and every component of U is open, then the component of U to which x belongs is a connected, open neighborhood of x contained in U.)

10.1. *Theorem.* Let X be a space and V an open subset of βX. Then

(a) $V \cap X$ is C^*-embedded in V; and

(b) $V \cap X$ is M-embedded in $V \cap \gamma X$.

Proof. (a) By Lemma 1.15 it is enough to prove that if $f \in C^*(V \cap X)$ and $p \in V$ then there is $\tilde{f} \in C^*((V \cap X) \cup \{p\})$ such that $f \subset \tilde{f}$. Let K be a compact neighborhood in βX of p such that $K \subset V$, choose $g \in C^*(\beta X)$ such that

$$g[K] = \{1\} \text{ and } g[X \backslash V] = \{0\},$$

and define $h : X \to R$ by

$$\begin{aligned} h(x) &= f(x) \cdot g(x) && \text{if } x \in V \cap X \\ &= 0 && \text{if } x \in X \backslash V. \end{aligned}$$

Since $h \in C^*(X)$ there is a continuous function $\bar{h} : \beta X \to R$ such that $h \subset \bar{h}$; we define

$$\tilde{f}(x) = f(x) \text{ for } x \in V \cap X, \text{ and}$$
$$\tilde{f}(p) = \bar{h}(p).$$

Then $(K \cap X) \cup \{p\}$ is a neighborhood of p in $(V \cap X) \cup \{p\}$ such that

$$\tilde{f} | (K \cap X) \cup \{p\} = \bar{h} | (K \cap X) \cup \{p\}.$$

It follows that $\tilde{f} \in C^*((V \cap X) \cup \{p\})$.

(b) From part (a) we have $V \cap \gamma X \subset V \subset \beta(V \cap X)$; it is enough to prove that $V \cap \gamma X \subset \gamma(V \cap X)$. If there is

$$q \in (V \cap \gamma X) \backslash \gamma(V \cap X)$$

then since $q \in \beta(V \cap X)$ and $\gamma(V \cap X)$ is topologically complete there is by Theorem 4.4(d) a locally finite cozero cover $\mathcal{U}$ of $V \cap X$ such that $U \in \mathcal{U}$ implies $q \notin c\ell_{\beta(V \cap X)} U$.

As in the proof of part (a) there is $Z \in Z(\beta X)$ and a cozero-set W of βX such that

$$q \in \mathrm{int}_{\beta X} Z \subset Z \subset W \subset c\ell_{\beta X} W \subset V.$$

We define

$$\mathcal{U}' = \{X \backslash Z\} \cup \{W \cap U : U \in \mathcal{U}\}$$

and we note that $\mathcal{U}'$ is a cover of X such that $U' \in \mathcal{U}'$ implies $q \notin c\ell_{\beta X} U'$. It follows from part (a) that the elements of $\mathcal{U}'$ are cozero sets of X. Finally $\mathcal{U}'$ is a locally finite cover of X, for if $x \in X \backslash V$ then $X \backslash c\ell_{\beta X} W$ is a neighborhood of x intersecting only one element of $\mathcal{U}'$, and if $x \in V$ then some neighborhood of x

in X intersects only finitely many elements of $\mathcal{U}$ (and hence of $\mathcal{U}'$). It follows from Theorem 3.2(d) that there are a metric space M and a continuous function $f : X \to M$ such that $\bar{f}(p) \in \beta M \setminus M$, and from Corollary 1.17(a) we have $p \in \gamma X \setminus X$. This contradiction completes the proof.

10.2. *Theorem.* For every space X, the following statements are equivalent.

(a) X is locally connected.

(b) γX is locally connected.

Proof. (a) ⇒ (b). Let U be an open neighborhood in γX of $p \in \gamma X$, and let V be an open subset of βX such that $U = V \cap \gamma X$. If $\{K_i : i \in I\}$ is a list of components of $V \cap X$ then since $\{K_i : i \in I\}$ is a locally finite co-zero cover of $V \cap X$ and (by Theorem 10.1) $p \in U \subset \gamma(V \cap X)$, there is (by Theorem 3.2(d) and Corollary 1.17(a)) $\bar{i} \in I$ such that $p \in c\ell_{\beta(V \cap X)} K_{\bar{i}}$. It is clear that $c\ell_{\beta(V \cap X)} K_{\bar{i}}$ is open in $\beta(V \cap X)$, so that $c\ell_U K_{\bar{i}}$ is an open, connected subset of U containing p.

(b) ⇒ (a). If U is an open neighborhood in X of $x \in X$ there is an open subset V of βX such that $U = V \cap X$, and since $x \in \gamma X$ there is a connected, open subset K of $V \cap \gamma X$ such that $x \in K$. It is clear from Theorem 10.1(a) that

$$K \cap X \subset K \subset \beta(K \cap X);$$

it follows that $K \cap X$ is an open, connected neighborhood of x such that $K \cap X \subset U$.

10.3. *Theorem.* If $p \in \beta X \setminus \upsilon X$ then βX is not locally connected at p.

Proof. Since $p \in \beta X \setminus \upsilon X$ there is by Theorem 2.1 a positive continuous function $f : X \to \mathbb{R}$ with no continuous real-valued extension to p. It is clear that if U is a neighborhood of p in βX then $f|U \cap X$ is an unbounded function. For every integer k such that $0 \le k \le 3$ we set

$$Z_k = \cup_{n<\omega} \{x \in X : 4n + k \le f(x) \le 4n + k + 1\}$$

and we note that $Z_k \in Z(X)$ for $k \le 3$ and that $\cup_{k\le 3} Z_k = X$. It follows that $p \in \cup_{k\le 3} c\ell_{\beta X} Z_k$; we assume without loss of generality that $p \in c\ell_{\beta X} Z_0$, so that $p \notin c\ell_{\beta X} Z_2$. If βX is locally connected at p there is a connected, open neighborhood U of p in βX such that $U \cap Z_2 = \emptyset$, and since

$$U \cap X \subset U \subset \beta(U \cap X)$$

(by Theorem 10.1(a)) the set $U \cap X$ is connected.

Now let $x \in U \cap X$ and choose $n < \omega$ and then $x' \in U \cap X$ such that

$$f(x) \le 4n < 4(n + 1) \le f(x').$$

Since $U \cap X$ is connected there is $z \in U \cap X$ such that $f(z) = 4n + 2$. This contradicts the fact that $U \cap Z_2 = \emptyset$.

10.4. *Lemma.* If f is a closed, continuous function from X onto Y and X is locally connected, then Y is locally connected.

Proof. We prove that if U is open in Y and K is a component of U, then K is open. We note first that if A is a connected subset of $f^{-1}(U)$ such that $A \cap f^{-1}(K) \ne \emptyset$, then $A \subset f^{-1}(K)$; indeed otherwise $f[A]$ is a connected subset of U intersecting both K and $U \backslash K$, contrary to the fact that K is a maximal connected subset of U. It follows that $f^{-1}(K)$ is open in $f^{-1}(U)$, hence in X, so that $X \backslash f^{-1}(K)$ is closed in X. From the facts that

$$K = Y \backslash f[X \backslash f^{-1}(K)]$$

and that f is a closed function it follows that K is open in X, as required.

10.5. *Theorem.* For every space X, the following statements are equivalent.

(a) X is locally connected and pseudocompact;

(b) βX is locally connected;

(c) every space in which X is dense is locally connected.

Proof. We note from Theorem 10.3 that if (a) or (b) holds then X is pseudocompact (so that $\beta X = \gamma X$ by the remark following Lemma 7.5). Thus the equivalence (a) $\Leftrightarrow$ (b) follows from Theorem 10.2. It is clear that (c) $\Rightarrow$ (b). Finally if βX is locally connected and X is dense in Y then since βY is the image of βX under a closed, continuous function the space βY is locally connected by Lemma 10.4 and hence Y is locally connected (by the implication (b) $\Rightarrow$ (a) applied to the space Y). Thus (b) $\Rightarrow$ (c) and the proof is complete.

§11. Some Miscellaneous Examples

We give three examples designed to answer questions suggested by some of the foregoing theorems. We prove first (Theorem 11.1) that not every closed Baire set is a zero-set. We prove next (Theorem 11.3), assuming the existence of an Ulam-measurable cardinal, that not every paracompact space X is the intersection of the locally compact, paracompact spaces between X and βX. And we conclude with several results responding to these questions raised by Theorem 9.10: Is every Baire set in a Lindelöf space a Lindelöf space? Is every Baire set in a paracompact space paracompact?

11.1. *Theorem.* Let R denote the real line and N the set of integers, and define

$$\Lambda = R \cup \{p \in \beta R : p \notin c\ell_{\beta R}\, N\}.$$

Then N is a closed Baire set of Λ, and $N \notin Z(\Lambda)$.

Proof. Since $N \subset R \subset \Lambda \subset \beta R$ and no accumulation point (in βR) of N is an element of Λ, N is closed in Λ. Further for $n \in N$ we have $\{n\} \in \mathcal{B}(\beta R)$, so that $\{n\} \in \mathcal{B}(\Lambda)$ and hence

$$N = \cup_{n<\omega} \{n\} \in \mathcal{B}(\Lambda).$$

Suppose finally that there is $f \in C^*(\Lambda)$ such that $N = Z(f)$, and for $n \in N$ choose $x_n \in R$ such that

$$0 < |n - x_n| < 1/n \quad\text{and}\quad 0 < f(x_n) < 1/n.$$

Since $\{x_n : n \in N\}$ and N are disjoint, closed subsets of R and R is a normal space, there is a continuous function $g : R \to [0,1]$ such that

$$g(x_n) = 0 \quad\text{and}\quad g(n) = 1 \quad\text{for}\quad n \in N;$$

then $\bar{g}[c\ell_{\beta R}\{x_n : n \in N\}] = \{0\}$ and $\bar{g}[c\ell_{\beta R}\, N] = \{1\}$. Since $\{x_n : n \in N\}$ is not compact there is

$$p \in c\ell_{\beta R}\{x_n : n \in N\} \backslash c\ell_{\beta R}\, N;$$

it is clear that $p \in \Lambda$ and $f(p) = 0$. This contradiction completes the proof.

The theorem just proved should be compared with §§9.15 - 9.17.

It is clear from Theorem 2.2 that for every space X we have

$$\upsilon X = \cap\{\text{coz } f : f \in C(\beta X),\ X \subset \text{coz } f\};$$

thus every realcompact space X is the intersection of the locally compact (*i.e.*, the open) σ-compact spaces between X and βX. It is natural to inquire whether the locally compact, paracompact, spaces are, in the same sense, "co-final" in the class of topologically complete spaces. Before answering this question (in the negative, by an example), we prove a theorem in the positive direction.

11.2. *Theorem.* If X is a space, then
$\gamma X = \cap\{Y : X \subset Y \subset \beta X,\ Y \text{ is paracompact and a } G_\delta \text{ in } \beta X\}$.

Proof. From Theorem 4.4 every paracompact space is topologically complete, so by Theorem 2.1 the indicated intersection contains γX. We assume without loss of generality that X is topologically complete and we prove that

$$X = \cap\{Y : X \subset Y \subset \beta X,\ Y \text{ is paracompact and a } G_\delta \text{ in } \beta X\}.$$

If $p \in \beta X \backslash X$ then by Theorems 4.4(d) and 3.2 there are a metric space M_p and a continuous function $f_p : X \to M_p$ such that for every compactification BM_p of M_p we have $\overline{f_p}(p) \in BM_p \backslash M_p$. Let $\overline{M_p}$ denote the metrizable completion of M_p and set $BM_p = \beta(\overline{M_p})$, and define

$$K_p = \overline{M_p} \backslash \{\overline{f_p}(p)\} \quad \text{and}$$

$$Y_p = \overline{f_p}^{-1}(K_p).$$

We note that K_p is completely metrizable by Lemma 3.4; we note also that since $M_p \subset K_p \subset BM_p$ and $\overline{f_p}[\beta X \backslash Y_p] = BM_p \backslash K_p$, the function $\overline{f_p}|Y_p$ is a perfect function from Y_p onto K_p by Lemma 9.2. It follows from Lemma 9.4((c) $\Rightarrow$ (a)) that Y_p is paracompact, and a G_δ in βY_p (*i.e.*, in βX).

Finally we have

$$X = \cap\{Y_p : p \in \beta X\backslash X\},$$

as required. The proof is complete.

We remark that if there is no Ulam-measurable cardinal and if X is a space, then $\gamma X = \upsilon X$ by Theorem 6.3 and hence γX is equal to the intersection of the open paracompact spaces (indeed, the open Lindelöf spaces) between X and βX.

Definition. Let α be an infinite cardinal, set $X = [0,1]^\alpha$, and define

$$H(\alpha) = \{p \in X : |\{\xi < \alpha : p_\xi \neq 0\}| \leq 1\}.$$

Let $\mathcal{O}$ be the element of $H(\alpha)$ such that $\mathcal{O}_\xi = 0$ for $\xi < \alpha$, and let ρ be the metric defined on the set $H(\alpha)$ as follows:

$\rho(p,p) = 0$ for $p \in H(\alpha)$;

if $p \in H(\alpha)$ and $p_\xi \neq 0$, then $\rho(p,\mathcal{O}) = \rho(\mathcal{O},p) = p_\xi$;

if $p, q \in H(\alpha)$ and there is $\xi < \alpha$ such that $p_\xi \neq 0$ and $q_\xi \neq 0$, then $\rho(p,q) = |p_\xi - q_\xi|$; and

if $p, q \in H(\alpha)$ and there are $\xi,\eta < \alpha$ such that $\xi \neq \eta$, $p_\xi \neq 0$ and $q_\eta \neq 0$, then $\rho(p,q) = p_\xi + q_\eta$.

Then $(H(\alpha),\rho)$ is the α-*hedgehog space*.

11.3. *Theorem.* Let α be an Ulam-measurable cardinal. Then the α-hedgehog space $H(\alpha)$ is paracompact and not realcompact, and if Y is a locally compact, paracompact space such that $H(\alpha) \subset Y \subset \beta H(\alpha)$, then

$$H(\alpha) \subset \upsilon H(\alpha) \subset Y.$$

Thus $H(\alpha)$ is not equal to the intersection of all locally compact, paracompact spaces between $H(\alpha)$ and $\beta H(\alpha)$.

Proof. It is clear that $H(\alpha)$ is a metric space of Ulam-measurable cardinality. Thus $H(\alpha)$ is paracompact, and (by Theorem 6.2) the space $H(\alpha)$ is not realcompact.

It is also clear, since any two elements of $H(\alpha)$ are elements of a subspace homeomorphic to $[0,1]$, that $H(\alpha)$ is a connected space.

Let Y be a locally compact, paracompact space such that $H(\alpha) \subset Y \subset \beta H(\alpha)$. By Lemma 9.9 there is a family $\{Y_i : i \in I\}$ of pairwise disjoint, open-and-closed, σ-compact subsets of Y such that $Y = \cup_{i \in I} Y_i$. For every $i \in I$ the space $Y_i \cap H(\alpha)$ is open-and-closed in $H(\alpha)$, and hence for every $i \in I$ either $Y_i \cap H(\alpha) = \emptyset$ or $Y_i \cap H(\alpha) = H(\alpha)$. Since $H(\alpha) \neq \emptyset$ there is $\bar{i} \in I$ such that $H(\alpha) \subset Y_{\bar{i}}$, and since $H(\alpha)$ is dense in Y we have

$$Y_i = \emptyset \quad \text{for} \quad i \in I \quad \text{and} \quad i \neq \bar{i};$$

hence $Y = Y_{\bar{i}}$. Since Y is σ-compact (and hence a Lindelöf space) and $H(\alpha) \subset Y$, we have $\upsilon H(\alpha) \subset Y$ by Corollary 2.4 and Theorem 2.1. The proof is complete.

In what follows we denote by I the interval $[0,1]$ and we set

$$J = \{p \in I : p \text{ is irrational}\} \text{ and}$$
$$Q = \{p \in I : p \text{ is rational}\}.$$

We denote the usual topology on I by T. If $A \subset I$ then T_A denotes the smallest topology on I such that $T \subset T_A$ and A is T_A-discrete -- *i.e.*, for $U \subset I$ we have $U \in T_A$ if and only if there are $T \in T$ and $B \subset A$ such that $U = T \cup B$.

11.4. *Lemma.* Let $A \subset I$.

(a) The space (I, T_A) is paracompact;

(b) if $A \subset X \subset I$ and $|A \cap K| \leq \omega$ for every T-compact subspace K of X, then $A \cup (I \backslash X)$, with the topology inherited from T_A, is a Lindelöf space.

Proof. We note first that (I, T_A) is a space. Since $T \subset T_A$ and (I, T) satisfies the Hausdorff separation axiom, also (I, T_A) satisfies that axiom. Further if $p \in U \in T_A$ then either there is $T \in T$ such that $p \in T \subset U$ or $p \in A$;

in either case there is a continuous function f from $(\mathcal{I},\mathcal{T}_A)$ to $[0,1]$ such that $f(p) = 0$ and $f[\mathcal{I}\backslash U] \subset \{1\}$. Thus $(\mathcal{I},\mathcal{T}_A)$ is a space.

(a) Let $\{U_i : i \in I\}$ be a cover of $\mathcal{I}$ by elements of $\mathcal{T}_A$, for $i \in I$ let $U_i = T_i \cup B_i$ with $T_i \in \mathcal{T}$ and $B_i \subset A$, and set $T = \bigcup_{i\in I} T_i$. Since $\mathcal{I}$ with the usual topology is paracompact (by Theorem 3.1(a)) there is a $\mathcal{T}$-open locally finite refinement $\mathcal{V}$ of $\{T_i : i \in I\}$. Then

$$\mathcal{V} \cup \{\{p\} : p \in \mathcal{I}\backslash\bigcup\mathcal{V}\}$$

is a $\mathcal{T}_A$-open locally finite refinement of $\mathcal{T}$.

(b) Let $\{S_i : i \in I\}$ be a cover of $A \cup (\mathcal{I}\backslash X)$ by (relatively) $\mathcal{T}_A$-open sets, for $i \in I$ let U_i be an element of $\mathcal{T}_A$ such that

$$S_i = U_i \cap [A \cup (\mathcal{I}\backslash X)],$$

let $U_i = T_i \cup B_i$ with $T_i \in \mathcal{T}$ and $B_i \subset A$, and define

$$T = \bigcup_{i\in I} T_i \quad\text{and}\quad K = \mathcal{I}\backslash T.$$

Since $\mathcal{I}$ with the usual topology is a Lindelöf space there is $J \subset I$ such that $|J| \le \omega$ and $\bigcup_{i\in I} T_i = \bigcup_{i\in J} T_i$, and since K with the usual topology is a compact subset of X we have $|A \cap K| \le \omega$ and there is $J' \subset I$ such that $|J'| \le \omega$ and $A \cap K \subset \bigcup_{i\in J'} B_i$. Then $\{S_i : i \in J \cup J'\}$ is a countable cover of $A \cup (\mathcal{I}\backslash X)$ by elements of $\{S_i : i \in I\}$.

We remark that there are $A \subset \mathcal{I}$ such that $(\mathcal{I},\mathcal{T}_A)$ is not a Lindelöf space. Let λ denote the usual Lebesgue measure for $\mathcal{I}$, let U be a $\mathcal{T}$-open subset of $\mathcal{I}$ such that $\mathcal{Q} \subset U$ and $\lambda U < 1$, and set $A = \mathcal{I}\backslash U$. Then (A is Lebesgue-measurable and) $\lambda A > 0$; it follows that $|A| \ge \omega^+$. It is clear that

$$\{U\} \cup \{\{p\} : p \in A\}$$

is a $\mathcal{T}_A$-open cover of $\mathcal{I}$ with no countable subcover.

11.5. *Theorem.* (a) If $A = \mathcal{J}$ and X denotes the set $\mathcal{I}$ with the topology $\mathcal{T}_A$, then $X \times \mathcal{J}$ is not normal.

(b) If A is a subset of $\mathcal{J}$ such that $|A| > \omega$ and

$|A \cap K| \leq \omega$ for every $\mathcal{T}$-compact $K \subset \mathcal{J}$, and if X denotes the set $A \cup \mathcal{Q}$ with the topology inherited from $\mathcal{T}_A$, then $X \times \mathcal{J}$ is not normal.

Proof. [In both (a) and (b), the product topology on $X \times \mathcal{J}$ is defined by the indicated topology on X and by the usual (metrizable) topology on $\mathcal{J}$].

We define

$$H = \{\langle p,p\rangle : p \in A\} \text{ and}$$
$$K = \mathcal{Q} \times \mathcal{J}$$

and we note that H and K are disjoint, closed subsets of $X \times \mathcal{J}$. To prove that $X \times \mathcal{J}$ is not normal it is enough to show that for every neighborhood W of H in $X \times \mathcal{J}$ there is $\langle q,p\rangle \in \mathcal{Q} \times \mathcal{J}$ such that every neighborhood in $X \times \mathcal{J}$ of $\langle q,p\rangle$ has non-empty intersection with W.

We define

$S_{1/n}(p) = \{p' \in \mathcal{J} : |p - p'| < 1/n\}$ for $p \in A$, $n < \omega$,

$A_n = \{p \in A : \{p\} \times S_{1/n}(p) \subset W\}$ for $n < \omega$, and

$B_n = c\ell_{(\mathcal{I},\mathcal{T})} A_n$ for $n < \omega$.

Since W is a neighborhood of H we have $A = \cup_{n<\omega} A_n$. We claim that there is $n < \omega$ such that $\mathcal{Q} \cap B_n \neq \emptyset$. If the claim fails then in (a) we have

$$\mathcal{J} = A = \cup_{n<\omega} A_n \subset \cup_{n<\omega} B_n \subset \mathcal{J}$$

and hence $\mathcal{J}$ is σ-compact, a contradiction; and in (b), using the fact that B_n is $\mathcal{T}$-compact for all $n < \omega$, from $A = \cup_{n<\omega} (A \cap B_n)$ we have

(*) $\quad |A| \leq \Sigma_{n<\omega} |A \cap B_n| \leq \omega \cdot \omega = \omega$,

a contradiction. The claim is proved: There are $n < \omega$ and $q \in \mathcal{Q}$ such that $q \in B_n$.

Let p be an element of $\mathcal{J}$ such that $|p - q| < 1/(2n)$ and let $U \times V$ be a neighborhood in $X \times \mathcal{J}$ of $\langle q,p\rangle$. We assume without loss of generality that $|q - x| < 1/(2n)$ for $x \in U$ and we prove that $(U \times V) \cap W \neq \emptyset$. Indeed there is $T \in \mathcal{T}$ such that $q \in T \subset U$, and since $q \in c\ell_{(\mathcal{I},\mathcal{T})} A_n$ there is

$$p' \in A_n \cap T \subset A_n \cap U.$$

Since $|p - p'| \le |p - q| + |q - p'| < 1/n$ and $p' \in A_n$ we have

$$\langle p', p\rangle \in W \cap (U \times V);$$

the proof is complete.

Lemmas 11.6 and 11.7 relate to the (possible) definition of a set A as in Theorem 11.5(a). The consequences of Theorem 11.5 for the families $\mathcal{B}(X)$ are given in Corollary 11.8 below.

For functions $f, g : \omega \to \omega$ we write

$$f \le g \quad \text{if} \quad f(n) \le g(n) \quad \text{for} \quad n < \omega.$$

It is clear that $\langle \omega^\omega, \le\rangle$ is a partially ordered set. As usual, a subset $\mathcal{S}$ of ω^ω is said to be *cofinal* (with respect to the order $\le$) if for $f \in \omega^\omega$ there is $g \in \mathcal{S}$ such that $f \le g$.

Definitions. Let X be a space.

(a) A *compact cover* of X is a family $\mathcal{K}$ of compact subsets of X such that $X = \cup\mathcal{K}$.

(b) The *compact-covering number* of X, denoted κX, is the least cardinal α such that there is a compact cover $\mathcal{K}$ of X with $|\mathcal{K}| = \alpha$.

11.6. *Lemma.* Let $\alpha \ge \omega$. The following statements are equivalent.

(a) There is a cofinal subset $\mathcal{S}$ of ω^ω such that $|\mathcal{S}| \le \alpha$;

(b) there is a compact cover $\mathcal{K}$ of $\mathcal{J}$ such that $|\mathcal{K}| \le \alpha$ and for every compact $K \subset \mathcal{J}$ there is $L \in \mathcal{K}$ such that $K \subset L$;

(c) $\kappa\mathcal{J} \le \alpha$.

Proof. It is well-known (see for example Kuratowski

[58] (page 88) or Sierpiński [56] (page 143)) that the space $\mathcal{J}$ is homeomorphic to the power space ω^ω; in what follows we identify these spaces. Further (for this proof only) for $n < \omega$ we set $[0,n] = \{k < \omega : 0 \le k \le n\}$.

For K a compact subset of ω^ω and $n < \omega$ we have

$$\sup \{f(n) : f \in K\} < \omega$$

and hence the function g_K given by

$$g_K(n) = \sup \{f(n) : f \in K\} \quad \text{for} \quad n < \omega$$

is a well-defined element of $\mathcal{J}$.

(a) $\Rightarrow$ (b). Let $\mathcal{S} = \{f_\xi : \xi < \alpha\}$ and define

$$K_\xi = \{g \in \omega^\omega : g \le f_\xi\} \quad \text{for} \quad \xi < \alpha.$$

Clearly K_ξ is a closed subset of ω^ω, and from $K_\xi \subset \prod_{n<\omega} [0,f_\xi(n)]$ it follows that K_ξ is compact; further for K a compact subset of ω^ω there is $\xi < \alpha$ such that $g_K \le f_\xi$ (and hence $K \subset K_\xi$). Thus the family $\mathcal{K} = \{K_\xi : \xi < \alpha\}$ is as required.

(b) $\Rightarrow$ (c) is obvious.

(c) $\Rightarrow$ (a). Let $\mathcal{K}$ be a compact cover of $\mathcal{J}$ such that $|\mathcal{K}| \le \alpha$. The family $\mathcal{S} = \{g_K : K \in \mathcal{K}\}$ is as required.

11.7. *Lemma.* There is $A \subset \mathcal{J}$ such that
$|A| = \kappa\mathcal{J}$ and
if K is a compact subset of $\mathcal{J}$ then $|A \cap K| < \kappa\mathcal{J}$.

Proof. Let $\{K_\xi : \xi < \kappa\mathcal{J}\}$ be a compact cover of $\mathcal{J}$. We assume without loss of generality, using Lemma 11.7((c) $\Rightarrow$ (b)), that if K is a compact subset of $\mathcal{J}$ then there is $\xi < \kappa\mathcal{J}$ such that $K \subset K_\xi$.

Recursively for $\xi < \kappa\mathcal{J}$ we choose

$$x_\xi \in \mathcal{J} \backslash \cup_{\zeta<\xi} (K_\zeta \cup \{x_\zeta\})$$

and we set $A = \{x_\xi : \xi < \alpha\}$. Clearly $|A| = \kappa\mathcal{J}$; and if K is a compact subset of $\mathcal{J}$ there is $\xi < \kappa\mathcal{J}$ such that $K \subset K_\xi$, and from

$$A \cap K_\xi \subset \{x_\zeta : \zeta \le \xi\}$$

we have $|A \cap K| \leq |A \cap K_\xi| \leq |\xi + 1| < \kappa\mathcal{J}$, as required. The proof is complete.

It is clear that $\kappa\mathcal{J} \leq 2^\omega$ and (since the space $\mathcal{J}$ is not σ-compact) that $\kappa\mathcal{J} \geq \omega^+$. Thus if the continuum hypothesis ($\omega^+ = 2^\omega$) is assumed we have $\kappa\mathcal{J} = \omega^+$ and it is possible to define a set A as in Theorem 11.5(b). For references to related, more delicate assumptions, see Remarks 11.10(b),(c),(d) below.

11.8. *Corollary.* (a) There are a paracompact space Y and $B \in \mathcal{B}(Y)$ such B is not paracompact.

(b) Assume $\omega^+ = 2^\omega$. There are a Lindelöf space Y and $B \in \mathcal{B}(Y)$ such that B is not a Lindelöf space.

Proof. By Lemma 11.4 and Theorem 11.5 there is a paracompact space X, which if $\omega^+ = 2^\omega$ may (according to the remark following Lemma 11.7) be taken to be a Lindelöf space, such that $X \times \mathcal{J}$ is not normal. We set $Y = X \times I$ and $B = X \times \mathcal{J}$. Then $B \in \mathcal{B}(Y)$ since $\mathcal{J} \in \mathcal{B}(I)$. An appeal to Lemma 4.2 and Corollary 4.5 completes the proof.

11.9. *Corollary.* (a) There is a countable family $\{X_n : n < \omega\}$ of spaces such that

$\prod_{n<k} X_n$ is paracompact for $k < \omega$, and
$\prod_{n<\omega} X_n$ is not normal.

(b) Assume $\omega^+ = 2^\omega$. There is a countable family $\{X_n : n < \omega\}$ of spaces such that

$\prod_{n<k} X_n$ is a Lindelöf space for $k < \omega$, and
$\prod_{n<\omega} X_n$ is not normal.

Proof. We use again the well-known fact (see for example Kuratowski [58] (page 88) or Sierpiński [56] (page 143)) that the space $\mathcal{J}$ is homeomorphic to ω^ω. Now in (a) and (b) let X_0 be the space X defined in parts (a) and (b) of Corollary 11.8, and set $X_n = \omega$ for $1 \leq n < \omega$. Since $\prod_{n<k} X_n$ is homeomorphic to the disjoint union of

countably many open-and-closed copies of X_0, the statements follow from Theorem 11.5.

11.10. *Remarks.* (a) It is apparently unknown if, without special set-theoretic assumptions, one can find a non-Lindelöf Baire set in a Lindelöf space. Similarly it is unknown if it is possible to define a countable family $\{X_n : n < \omega\}$ of spaces such that $\prod_{n<k} X_n$ is a Lindelöf space for $k < \omega$ but $\prod_{n<\omega} X_n$ is not a Lindelöf space. We have proved above that the assumption $\kappa\mathcal{J} = \omega^+$ is enough to settle both these questions in the affirmative and we have noted that the assumption $\omega^+ = 2^\omega$ implies $\kappa\mathcal{J} = \omega^+$.

(b) Let $A, B \subset I$. Then B is *concentrated about* A (cf. Besicovitch [34]) if $|B\backslash U| \leq \omega$ for every open subset U of I such that $A \subset U$. The following result has been communicated by R. Dan Mauldin (letter to the authors, September, 1974) (cf. Theorem 11.5(b)).

Theorem. There is an uncountable subset A of $\mathcal{J}$ such that $|A \cap K| \leq \omega$ for every compact subset K of $\mathcal{J}$ if and only if some uncountable subset of I is concentrated about a countable set.

(c) A subset A of I is called a *Sierpiński* set [a *Lusin* set] if A is an uncountable set such that $|A \cap K| \leq \omega$ for every set K of Lebesgue measure zero [for every nowhere dense set K]. We note that if A is a Lusin set then $|A \cap K| \leq \omega$ for every compact subset K of $\mathcal{J}$ (cf. Theorem 11.5(b)). An argument similar to that of Lemma 11.7 shows that if the continuum hypothesis $(\omega^+ = 2^\omega)$ is assumed then there is a Sierpiński set A and there is a Lusin set B such that $|A| = |B| = 2^\omega$ (cf. Mahlo [13] (Aufgabe 5)); the converse has been established by Rothberger [38].

(d) The following hypothesis, (a topological version of) Martin's axiom, is used most frequently in conjunction with the denial of the continuum hypothesis. We note that if the continuum hypothesis $(\omega^+ = 2^\omega)$ is assumed, then Martin's axiom is a consequence of the classical Baire category theorem for compact spaces.

Martin's Axiom. If X is a compact Hausdorff space with no uncountable family of pairwise disjoint non-empty open subsets, and if $X = \cup_{i \in I} A_i$ with A_i nowhere dense in X, then $|I| \geq 2^\omega$.

Statements (1), (2) and (3) below are easily established using Martin's axiom and are to be compared with Lemma 11.7, Theorem 11.5(b), and Corollary 11.8(b) respectively. In (2), for α a cardinal we denote (as usual) by $cf(\alpha)$ the least cardinal β such that there is a family $\{\alpha_\xi : \xi < \beta\}$ of cardinals with

$$\alpha_\xi < \alpha \text{ for } \xi < \beta \text{ and}$$
$$\Sigma_{\xi<\beta}\, \alpha_\xi = \alpha.$$

Recall that for $\alpha \geq \omega$ a space X is α-*compact* if for every open cover $\mathcal{U}$ of X there is a cover $\mathcal{V}$ of X such that $\mathcal{V} \subset \mathcal{U}$ and $|\mathcal{V}| < \alpha$.

Theorem. Assume Martin's axiom.

(1) There is $A \subset \mathcal{J}$ such that $|A| = 2^\omega$ and $|A \cap K| < 2^\omega$ for every compact subset K of $\mathcal{J}$.

(2) Let $\omega < cf(\alpha) \leq \alpha \leq 2^\omega$ and let A be a subset of $\mathcal{I}$ such that $|A| = \alpha$ and $|A \cap K| < \alpha$ for every $\mathcal{T}$-compact $K \subset \mathcal{J}$. If X denotes the set $A \cup \mathcal{Q}$ with the topology inherited from $\mathcal{T}_A$, then $X \times \mathcal{J}$ is not normal.

(3) There are a 2^ω-compact space Y and $B \in \mathcal{B}(Y)$ such that B is not normal.

In the proof of (2), statement (*) of (the proof of) Theorem 11.5(b) is replaced by

$$\alpha = |A| \leq \Sigma_{n<\omega} |A \cap B_n|,$$

contradicting the assumption $\omega < cf(\alpha)$. The remaining details of the proof of the theorem are omitted.

(e) The cardinal number κJ has been studied, and compared with other cardinals, by Hechler [73], [74], [75]. It is remarked by Hechler [75] that $\kappa J = \omega^+$ in the models of Solovay [70] formed by the adjunction of random reals and that $\kappa J = 2^\omega$ in the models of Cohen [66] using generic reals. It is clear that if Martin's axiom is assumed then $\kappa J = 2^\omega$. Indeed from Martin's axiom it follows that if $A \subset \omega^\omega$ and $|A| < 2^\omega$ then there is $f \in \omega^\omega$ such that $g \in A$ implies $|\{n < \omega : f(n) \leq g(n)\}| < \omega$; for this and related results concerning ω^ω see Kunen [68], Martin and Solovay [70], Booth [71] and Tall [74].

Notes and References

§1. $\bar{P}$-reflections

Theorem 1.2. Tychonoff [30]; Engelking [68] (Theorem 2.3.8).

Corollary 1.3. Tychonoff [30]; M. H. Stone [37] (Theorems 79 and 88); Čech [37] (Theorem, page 831).

Lemma 1.4. Engelking [68] (Theorem 1.5.2).

Lemma 1.5. Čech [37]; Gillman and Jerison [60] (Lemma 6.11).

The principal references for 1.8-1.14 are van der Slot [66] and Herrlich [67]. See also Freyd [64] (Exercise 3J, "The Adjoint Functor Theorem"), Isbell [64] (I.25-27), Kennison [65], Herrlich and van der Slot [67], Herrlich [69], and Franklin [71].

Lemma 1.15. Bourbaki and Dieudonné [39]; McDowell [58] (Proposition 3.2).

§2. Realcompact Spaces and Topologically Complete Spaces

Realcompact spaces were defined and investigated by Hewitt [48] (§II.7) and independently by Nachbin (unpublished); see Hewitt [53], Nachbin [52], [54], and Gillman and Jerison [60] (Chapter 8; Notes to Chapter 8).

Responding to a question posed in the fundamental memoire of Weil [37], Dieudonné [39] (Theorem, page 285) proved that a uniformisable space has a compatible complete uniformity if and only if it is (in our terminology) topologically complete. These spaces are studied by Kelley [55] (Chapter 6), Gillman and Jerison [60] (Chapter 15), and Isbell [64] (I.10-22).

Theorem 2.2. Shirota [52] (Theorem 1); Hewitt [48] (Theorem 60).

Theorem 2.3. Smirnov [51].

The statement that every open cover of a Lindelöf space has a star-finite open refinement, closely related to Corollary 2.4, is given by Morita [48] (Theorem 10).

Lemma 2.5. Ulam [30]; Tarski [38] (page 153).

Theorem 2.6. Mackey [44].

§3. Metric Spaces

Paracompact spaces were introduced by Dieudonné [44], who proved Theorem 3.1 for separable metric spaces. The general case is due to A. H. Stone [48]. Our proof is from M. E. Rudin [69]; see also Ornstein [69].

Parts of Theorem 3.2 are given by Kodaira [41] (§5), Kakutani and Kodaira [44], Halmos [50] (Theorem 51.D), Michael [53], and Tamano [60], [62].

Our proof of Lemma 3.3 follows Kuratowski [58] (§31,I).

Lemma 3.4. Alexandroff [24].

Lemma 3.5 and Theorem 3.6. Čech [37] (§III). The implication (c) ⇒ (a) of Theorem 3.6 is given also by Dieudonné [39] (Theorem, page 283).

We note that in the terminology of Čech [37] a space is "topologically complete" if it is a G_δ in one of its compactifications.

§4. X as a Subset of βX; Locally Finite Covers

Lemma 4.1. See Morita [48] (Theorem 3) and Shirota [52] (Remark, page 23).

Lemma 4.2(a). Morita [48] (Theorem 10).

Lemma 4.2(b). Dieudonné [44].

Lemma 4.3. Michael [53] (Proposition 2). Statement (*) in the proof of Lemma 4.3 is from Dieudonné [44] (Théorème 6).

Theorem 4.4(a). Smirnov [51] and Tamano [62] (Theorem 2.7).

Theorem 4.4(b). Hewitt [48] (Theorem 50); Tamano [62] (Theorem 2.5).

Theorem 4.4(c). Tamano [60] (Corollary, page 1046); Tamano [62] (Theorem 2.8).

Theorem 4.4(d). Kac [54]; Tamano [62] (Theorem 2.6).

For further characterizations of realcompact and paracompact spaces using partitions of unity, see De Marco and Wilson [71].

A direct proof of Corollary 4.5(c) is given by Dieudonné [44] (Théorème 5). Using perfect functions, Henriksen and Isbell [58] (Theorem 2.2 and page 93) show for many properties P (including those of Corollary 4.5) that the product of a compact space and a space with P has P.

§5. Tamano's Theorem

The principal result of this Section, Theorem 5.3 ((a) ⇔ (c)), is from Tamano [60] (Theorem 2). Proofs of the other equivalences of Theorem 5.3 are given by Tamano [62] (Theorem 3.1), Corson [62], and Morita [62] (Theorem 2.7).

See also Corson [58].

Lemma 5.1. Mrówka [59] (Theorem II(ii)); Frolík [60a] (Lemma 1.1).

Theorem 5.4. Tamano [62] (Theorem 3.7).

§6. The Katětov-Shirota Theorem

Theorem 6.3 is due in its full generality to Shirota [52] (§3) and for paracompact spaces to Katětov [51a] (Theorem 3). The embedding technique of the second proof has been used by Glicksberg [59] (Lemma 1) and Comfort and Negrepontis [66] (Theorem 2.8).

A proof of Katětov's Theorem appears in Mrówka [64].

The proof of Theorem 6.2((c) ⇒ (a)) is taken (with modifications) from Gillman and Jerison [60] (§§15.18-15.19).

For more detailed comments about the unprovability in ZFC of Axiom (U) (mentioned in Remark 6.5), the reader is referred to Schoenfield [67] (§9.10 and Problem 9.14).

§7. On the Relations $P(X \times Y) = PX \times PY$

Lemma 7.1. Comfort and Negrepontis [66] (Theorem 5.2).

Theorem 7.2. Comfort [74]; Comfort and Herrlich [*].

Corollary 7.4(a). Pupier [69] (Théorème 2.1); Morita [70] (Theorem 5.1).

Corollary 7.4(b). Husek [70] (Theorem 2).

For additional results concerning the relation $\upsilon(X \times Y) = \upsilon X \times \upsilon Y$, see Comfort [67], [68]; Hager [72]; McArthur [70]; Husek [72]; and Blair and Hager [*]. Concerning $\gamma(X \times Y) = \gamma X \times \gamma Y$, see Isiwata [71], Buchwalter

[71], and Pupier [72].

Pseudocompact spaces were introduced by Hewitt [48].

Lemma 7.5. Mardesić and Papić [55]; Bagley, Connell and McKnight [58].

Theorem 7.6. Glicksberg [59]; Frolík [60a]; Henriksen and Isbell [57b].

§8. Absolute (Separable, Metrizable) Borel Spaces

Lemma 8.1 is from Negrepontis [65] (Lemma 1.14); the proof given here is due to J. H. B. Kemperman. See also Christensen [74] (2.6: Lemma 3).

Lemma 8.3 and Corollary 8.4. Kuratowski [58] (§26,III).

A family $\{A_\xi : \xi < 2^\omega\}$ of subsets of ω such that $|A_\xi \cap A_\zeta| < \omega$ for $\xi < \zeta < 2^\omega$, as in Theorem 8.6, has been defined by Sierpiński [28]; see Tarski [28] for generalizations. The space X of Theorem 8.6 was defined (for a purpose different from ours) by Katětov [50] (Example 1); see also Gillman and Jerison [60] (Exercise 6Q).

Lemma 8.8. Urysohn [25].

Lemma 8.9. Lavrentieff [24] (§1).

Theorem 8.10. Mazurkiewicz [16]; Sierpiński [26].

Theorem 8.11. Lavrentieff [24] (§II, Théorèmes I and II).

§9. Topological Properties of Baire Sets

Lemma 9.1. Kodaira [41] (§5); Halmos [50] (Theorem 51.D); Negrepontis [67a]; Comfort [70].

Our principal reference for perfect functions and their properties is Henriksen and Isbell [58]. Lemma 9.2 and (most of) Corollary 9.3 appear as Lemma 1.5 and Theorem 2.2 respectively in Henriksen and Isbell [58]. For early work on the subject see Vaǐnšteǐn [47], [52], Leray [50] and Whyburn [50]. P. S. Alexandroff [60] (§5, especially footnote 1 of page 55) gives a historical perspective.

Theorem 9.4 is from Frolík [60b] (Lemma 1, Theorem 3); our proof is different from his.

Corollary 9.5. Frolík [60b]; Engelking [66] (Remark 2).

Theorem 9.6. Frolík [64] (Theorem 1); the proof given here appears in Negrepontis [67a] (Theorem 2.1).

Lemma 9.7. Henriksen, Isbell and Johnson [61] (Lemma 2.2).

Lemma 9.9. Bourbaki [65] (§9.10, Théorème 5).

Theorem 9.10(a). Frolík [62] (§2); Sion [60] (Theorem 2.3); Henriksen, Isbell and Johnson [61] (Corollary 2.3); Negrepontis [67a] (Theorem 2.6).

Theorem 9.10(b). Lorch [63] (Theorem 8); Negrepontis [67b] (Theorem 3.8).

Theorem 9.10(c). Negrepontis [67b] (Theorem 4.2).

The first (implicit) mention known to the authors of the concept of a Z-embedded subset is in Isbell [58] (Remark 1.31). Lemma 9.11(a) is due to Meyer Jerison (see Henriksen and Johnson [61] (Lemma 5.3)).

Theorem 9.12. Frolík [70] (Theorem 15). See Willard [66] for the "Borel" analogue.

Lemma 9.13. Frolík [60b] (Lemma 2).

Theorem 9.14(a). Engelking [68] (Theorem 3.8.5).

Theorem 9.14(b). Hager [69] (Remark 2.3(i)).

Theorem 9.14(c). Frolík [60b] (Theorem 1).

Theorem 9.14(d). Engelking [66] (Remark 2).

Theorem 9.14(e). Negrepontis [67a] (Lemma 2.9).

It is shown by Halmos [50] (Theorem 51.D) that every compact Baire set in a space X is a zero-set of X. Theorem 9.15 is from Comfort [70] (Theorem 1.2).

Corollary 9.16(a),(c). Ross and Stromberg [65] (Theorem 1.3).

Theorem 9.17. Negrepontis [67b] (Lemma 5.1).

§10. Local Connectedness in βX

The implication (b) $\Rightarrow$ (a) of Theorem 10.5 is from Banaschewski [56]. All the results of §10 appear in Henriksen and Isbell [57a]. For an alternative proof of Theorem 10.5((a) $\Leftrightarrow$ (b)), see Wulbert [69] (Theorem 2).

Lemma 10.4. Whyburn [52].

It has been shown by Wilder [49] (Corollary IV.2.2) that if Y is a locally compact, connected space then $\{p \in Y : Y$ is not locally connected at $p\}$, if not empty, contains a continuum (*i.e.*, a compact, connected set C such that $|C| > 1$). The following consequence of Wilder's result should be contrasted with Theorem 10.5: If X is locally compact, connected and locally connected, then the one-point compactification of X is locally connected.

§11. Some Miscellaneous Examples

Theorem 11.1 is from Ross and Stromberg [65] (Example 3.1). The space Λ had been defined (for a different purpose) by Katětov [51b] (Example, page 88); see Gillman and Jerison [60] (Exercise 6P).

The α-hedgehog space has been defined by Urysohn [27] and Schmidt [32]. See Engelking [68] (4.1 (Example 3) and Theorem 4.4.7) for a general discussion, and Comfort and Hager [*] for generalizations.

Topologies of the form T_A as defined in the paragraph preceding Lemma 11.4 were apparently first defined by Hanner [51] (proof of Theorem 4.2).

Lemma 11.4, Theorem 11.5. Michael [63].

Lemma 11.6((a) ⇒ (b)) is due to Hechler and Mrówka (see Hechler [74] (page 212) and Hechler [75] (Theorem 1)).

Corollary 11.8 is from Negrepontis [66], where a general result concerning the so-called compact properties is established using continuous pseudometrics.

Corollary 11.9. Michael [71] (Corollary 3.3).

The question in (the second part of) Remark 11.10(a) has been asked by Michael [70] and [71] (§7.1).

Martin's axiom (cf. Remark 11.10(d)) was introduced by Martin and Solovay [70] and Kunen [68] (§11). For examples of some of its uses in topology, see M. E. Rudin [75], Tall [74], [*], and Shinoda [73].

Bibliography

P. S. Alexandroff

[24] *Sur les ensembles de la première classe et les espaces abstraits,* Comptes Rendus Acad. Sci. (Paris) 178 (1924), 185-187.

[60] *Some results in the theory of topological spaces, obtained within the last twenty-five years.* In: Russian Math. Surveys 15 No. 2, pp. 23-83. American Mathematical Society, Providence, 1960.

R. W. Bagley, E. H. Connell and J. D. McKnight Jr.

[58] *On properties characterizing pseudo-compact spaces,* Proc. Amer. Math. Soc. 9 (1958), 500-506.

Bernhard Banaschewski

[56] *Local connectedness of extension spaces,* Canadian J. Math. 8 (1956), 395-398.

A. S. Besicovitch

[34] *Concentrated and rarified sets of points,* Acta Math. 62 (1934), 289-300.

Robert Blair and Anthony W. Hager

[*] *Notes on the Hewitt realcompactification of a product,* General Topology and its Applications, to appear.

David Booth

[71] *Generic covers and dimension,* Duke Math. J. 38 (1971), 667-670.

N. Bourbaki

[65] Topologie Générale, Chapter 1. Actualités Scientifiques et Industrielles, Quatrième Edition. Hermann, Paris, 1965.

N. Bourbaki and J. Dieudonné

[39] *Note de tératopologie II.* Revue Scientifique 77 (1939), 180-181.

Henri Buchwalter

[71] *Produit topologique, produit tensoriel et* c*-repletion.* In: Proc. Colloque (1971) International d'Analyse Fonctionelle de Bordeaux, pp. 1-26. Univ. Claude-Bernard (Lyon), multilith publication.

Eduard Čech

[37] *On bicompact spaces,* Annals of Math. 38 (1937), 823-844.

J. P. R. Christensen

[74] Topology and Borel Structure. North-Holland Mathematics Studies 10. North-Holland Publishing Co., Amsterdam, 1974.

Paul J. Cohen

[66] Set Theory and the Continuum Hypothesis. Benjamin, New York, 1966.

W. W. Comfort

[67] *Locally compact realcompactifications.* In: General Topology and its Relations to Modern Analysis and Algebra II. Proceedings of the Second (1966) Prague Topological Symposium, pp. 95-100. Academia, Prague, 1967.

[68] *On the Hewitt realcompactification of a product space,* Trans. Amer. Math. Soc. 131 (1968), 107-118.

[70] *Closed Baire sets are (sometimes) zero-sets,* Proc. Amer. Math. Soc. 25 (1970), 870-875.

[74] Review of Isiwata [71], Math. Reviews 47 #7697 (1974), 1339.

W. W. Comfort and Anthony W. Hager

[*] *Metric spaces without large closed discrete sets,* manuscript.

W. W. Comfort and H. Herrlich

[*] *On the relations* $P(X \times Y) = PX \times PY$, manuscript.

W. W. Comfort and S. Negrepontis

[66] *Extending continuous functions on* $X \times Y$ *to subsets of* $\beta X \times \beta Y$, Fund. Math. 59 (1966), 1-12.

[74] The Theory of Ultrafilters. Grundlehren der Math. Band 211. Springer-Verlag, Heidelberg, 1974.

H. H. Corson

[58] *The determination of paracompactness by uniformities*, American J. Math. 80 (1958), 185-190.

[62] Review of Tamano [60], Math. Reviews 23 #A2186 (1962), 407.

G. De Marco and R. G. Wilson

[71] *Realcompactness and partitions of unity*, Proc. Amer. Math. Soc. 30 (1971), 189-194.

J. Dieudonné

[39] *Sur les espaces uniformes complets*, Annales de l'école normale supérieure, 3me série, 56 (1939), 276-291.

[44] *Une généralisation des espaces compacts*, J. de Math. Pures et Appl. 23 (1944), 65-76.

R. Engelking

[66] *On functions defined on Cartesian products*, Fund. Math. 59 (1966), 221-231.

[68] Outline of General Topology. North-Holland Publishing Co., Amsterdam, 1968.

S. P. Franklin

[71] *On epi-reflective hulls*, General Topology and its Applications 1 (1971), 29-31.

Peter Freyd

[64] Abelian Categories. Harper and Row, New York, 1964.

Zdeněk Frolík

[60a] *The topological product of two pseudocompact spaces*,

Czech. Math. J. 10 (85) (1960), 339-348.

[60b] *On the topological product of paracompact spaces,* Bull. Acad. Polon. Sci. Sér. Sci. Math. Astr. Phys. 8 (1960), 747-750.

[62] *A contribution to the descriptive theory of sets and spaces.* In: General Topology and its Relations to Modern Analysis and Algebra, Proc. (1961) Prague Topological Symposium, pp. 157-173. Academic Press, New York, 1962.

[64] *On coanalytic and bianalytic spaces,* Bull. Acad. Polon. Sci. Sér. Sci. Math. Astr. Phys. 12 (1964), 527-530.

[70] *Absolute Borel sets and Souslin sets,* Pacific J. Math. 32 (1970), 663-683.

Leonard Gillman and Meyer Jerison

[60] Rings of Continuous Functions. D. Van Nostrand Co., Inc., Princeton, New Jersey, 1960.

Irving Glicksberg

[59] *Stone-Čech compactifications of products,* Trans. Amer. Math. Soc. 90 (1959), 369-382.

Anthony W. Hager

[69] *Approximation of real continuous functions on Lindelöf spaces,* Proc. Amer. Math. Soc. 22 (1969), 156-163.

[72] *Uniformities on a product,* Canadian J. Math. 24 (1972), 379-389.

Paul R. Halmos

[50] Measure Theory. D. Van Nostrand Co., Inc., New York, 1950.

Olof Hanner

[51] *Solid spaces and absolute retracts,* Arkiv för Mat. 1 (1951), 375-382.

Stephen H. Hechler

[73] *Independence results concerning the number of nowhere dense sets required to cover the real line,* Acta Math.

Acad. Sci. Hungar. 24 (1973), 27-32.

[74] *A dozen small uncountable cardinals.* In: TOPO 72 - General Topology and its Applications, Lecture Notes in Mathematics #378, pp. 207-218. Springer-Verlag, Heidelberg, 1974.

[75] *On a ubiquitous cardinal,* Proc. Amer. Math. Soc. (1975), to appear.

Melvin Henriksen and J. R. Isbell

[57a] *Local connectedness in the Stone-Čech compactification,* Illinois J. Math 1 (1957), 574-582.

[57b] *On the Stone-Čech compactification of a product of two spaces,* Bull. Amer. Math. Soc. 63 (1957), 145-146 (Abstract).

[58] *Some properties of compactifications,* Duke Math. J. 25 (1958), 83-105.

M. Henriksen, J. R. Isbell and D. G. Johnson

[61] *Residue class fields of lattice-ordered algebras,* Fund. Math. 50 (1961), 107-117.

[61] M. Henriksen and D. G. Johnson
On the structure of a class of archimedean lattice-ordered algebras, Fund. Math. 50 (1961), 73-94.

Horst Herrlich

[67] *𝕰-kompakte Räume,* Math. Zeitschrift 96 (1967), 228-255.

[69] *On the concept of reflections in general topology.* In: Contributions to Extension Theory of Topological Structures, Proc. (1967) Berlin Symposium, pp. 105-114. VEB Deutscher Verlag der Wissenschaften, Berlin, 1969.

H. Herrlich and J. van der Slot

[67] *Properties which are closely related to compactness,* Indag. Math. 29 (1967), 524-529.

Edwin Hewitt

[48] *Rings of real-valued continuous functions I,* Trans. Amer. Math. Soc. 64 (1948), 45-99.

[53] Review of Shirota [52], Math. Reviews 14 (1953), 395.

Miroslav Hušek

[70] *The Hewitt realcompactification of a product*, Comment. Math. Univ. Carolinae 11 (1970), 393-395.

[72] *Realcompactness of function spaces and* $\upsilon(P \times Q)$, General Topology and its Applications 2 (1972), 165-179.

J. R. Isbell

[58] *Algebras of uniformly continuous functions*, Annals of Math. (2) 68 (1958), 96-125.

[64] Uniform Spaces. Math. Surveys No. 12. American Mathematical Society, Providence, R.I., 1964.

Takesi Isiwata

[71] *Topological completions and realcompactifications*, Proc. Japan Acad. 47 (1971), 941-946.

G. I. Kac

[54] *Topological spaces in which one may introduce a complete uniform structure*, Doklady Akad. Nauk SSSR (N.S.) 99 (1954), 897-900 [in Russian].

Shizuo Kakutani and Kunihiko Kodaira

[44] *Über das Haarsche Mass in der lokal bicompacten Gruppe*, Proc. Imp. Acad. Tokyo 20 (1944), 444-450.

Miroslav Katětov

[50] *On nearly discrete spaces*, Časopis Pěstováni Mat. Fys. 75 (1950), 69-78.

[51a] *Measures in fully normal spaces*, Fund. Math. 38 (1951), 73-84.

[51b] *On real-valued functions in topological spaces*, Fund. Math. 38 (1951), 85-91 and 40 (1953), 203-205.

John L. Kelley

[55] General Topology. D. Van Nostrand Co., Inc., New York, 1955.

J. F. Kennison

[65] *Reflective functors in general topology and elsewhere*, Trans. Amer. Math. Soc. 118 (1965), 303-315.

Kunihiko Kodaira

[41] *Über die Beziehung den Massen und Topologien in einer Gruppe*, Proc. Phys.-Math. Soc. Japan 23 (1941), 67-119.

Kenneth Kunen

[68] *Inaccessibility properties of cardinals.* Doctoral dissertation. Stanford University, 1968.

Casimir Kuratowski

[58] Topologie Volume I, 4me édition. Polska Akademia Nauk, Warszawa, 1958.

M. M. Lavrentieff

[24] *Contribution à la théorie des ensembles homéomorphes*, Fund. Math. 6 (1924), 149-160.

Jean Leray

[50] *L'anneau spectral et l'anneau filtré d'homologie d'un espace localement compact et d'une application continue*, J. de Math. Pures et Appl. 29 (1950), 1-80.

E. R. Lorch

[63] *Compactification, Baire functions and Daniell integration*, Acta Sci. Math. Szeged 24 (1963), 204-218.

George W. Mackey

[44] *Equivalence of a problem in measure theory to a problem in the theory of vector lattices*, Bull. Amer. Math. Soc. 50 (1944), 719-722.

P. Mahlo

[13] *Über Teilmengen des Kontinuums von dessen Mächtigkeit*, Sitzungsberichte Leipzig 65 (1913), 283-315.

S. Mardesić and P. Papić

[55] *Sur les espaces dont toute transformation réelle*

continue est bornée, Hrvatsko Prirod. Drustvo. Glasnik Mat.-Fiz. Astr. Ser. II 10 (1955), 225-232.

D. A. Martin and R. M. Solovay

[70] *Internal Cohen extensions*, Annals of Math. Logic 2 (1970), 143-178.

S. Mazurkiewicz

[16] *Über Borelsche Mengen*, Bull. Acad. Sci. Cracovie (1916), 490-494.

William G. McArthur

[70] *Hewitt realcompactifications of products*, Canadian J. Math. 22 (1970), 645-656.

Robert H. McDowell

[58] *Extension of functions from dense subsets*, Duke Math. J. 25 (1958), 297-304.

Ernest Michael

[53] *A note on paracompact spaces*, Proc. Amer. Math. Soc. 4 (1953), 831-838.

[63] *The product of a normal space and a metric space need not be normal*, Bull. Amer. Math. Soc. 69 (1963), 375-376.

[70] *Paracompactness and the Lindelöf property for* X^n *and* X^ω. In: Proc. 1970 Washington State Univ. Conference on General Topology, pp. 11-12. Washington State University, Pullman, 1970.

[71] *Paracompactness and the Lindelöf property in finite and countable cartesian products*, Compositio Math. 23 (1971), 199-214.

Kiiti Morita

[48] *Star-finite coverings and the star-finite property*, Math. Japonicae 1 (1948), 60-68.

[62] *Paracompactness and product spaces*, Fund. Math. 50 (1962), 223-236.

[70] *Topological completions and* M-*spaces*, Sci. Reports

Tokyo Kyoiku Daigaku 10 (1970), 271-288.

S. Mrówka

[59] *Compactness and product spaces*, Colloquium Math. 7 (1959), 19-22.

[64] *An elementary proof of Katětov's theorem concerning Q-spaces*, Michigan Math. J. 11 (1964), 61-63.

Leopoldo Nachbin

[52] *On the continuity of positive linear transformations.* In: Proc. 1950 International Congress of Mathematicians, pp. 464-465. American Mathematical Society, Providence, R.I., 1952.

[54] *Topological vector spaces of continuous functions*, Proc. Nat. Acad. Sci. U.S.A. 40 (1954), 471-474.

S. Negrepontis

[65] *A homology theory for realcompact spaces.* Doctoral dissertation. University of Rochester, 1965.

[66] *Baire sets in paracompact spaces*, 1966, manuscript.

[67a] *Absolute Baire sets*, Proc. Amer. Math. Soc. 18 (1967), 691-694.

[67b] *Baire sets in topological spaces*, Archiv der Math. 18 (1967), 603-608.

N. Noble

[69] *Ascoli theorems and the exponential map*, Trans. Amer. Math. Soc. 143 (1969), 393-411.

Donald Ornstein

[69] *A new proof of the paracompactness of metric spaces*, Proc. Amer. Math. Soc. 21 (1969), 341-342.

René Pupier

[69] *La completion universelle d'un produit d'espaces complètement reguliers*, Publication T. 6.-2 du Département de Mathématiques du Faculté des Sciences de Lyon, 1969.

[72] *Le problème de la completion topologique d'un produit*,

Comptes Rendus Acad. Sci. (Paris) 274 (1972), 1884-1887.

Kenneth A. Ross and Karl R. Stromberg

[65] *Baire sets and Baire measures*, Arkiv för Mat. 6 (1965), 151-160.

Fritz Rothberger

[38] *Eine Äquivalenz zwischen der Kontinuumhypothese und der Existenz der Lusinschen und Sierpińskischen Mengen*, Fund. Math. 30 (1938), 215-217.

Mary Ellen Rudin

[69] *A new proof that metric spaces are paracompact*, Proc. Amer. Math. Soc. 20 (1969), 603.

[75] Lectures in Set Theoretic Topology. Regional Conference Series in Mathematics No. 23. American Mathematical Society, Providence, R.I., 1975.

Friedrich Karl Schmidt

[32] *Über die Dichte metrischer Räume*, Math. Annalen 106 (1932), 457-472.

Joseph R. Schoenfield

[67] Mathematical Logic. Addison-Wesley Publishing Co., Reading, Massachusetts, 1967.

Juichi Shinoda

[73] *Some consequences of Martin's axiom and the negation of the continuum hypothesis*, Nagoya Math. J. 49 (1973), 117-125.

Taira Shirota

[52] *A class of topological spaces*, Osaka Math. J. 4 (1952), 23-40.

W. Sierpiński

[26] *Sur l'invariance topologique des ensembles* G_δ, Fund. Math. 8 (1926), 135-136.

[28] *Sur une décomposition d'ensembles*, Monatshefte für Math. Phys. 35 (1928), 239-242.

Maurice Sion

[60] *Topological and measure theoretic properties of analytic sets*, Proc. Amer. Math. Soc. 11 (1960), 769-776.

J. van der Slot

[66] *Universal topological properties*, ZW 1966-011. Math. Centrum., Amsterdam, 1966.

Yuri M. Smirnov

[51] *On normally disposed sets of normal spaces*, Mat. Sbornik N. S. 29 (71) (1951), 173-176 [in Russian].

A. H. Stone

[48] *Paracompactness and product spaces*, Bull. Amer. Math. Soc. 54 (1948), 977-982.

M. H. Stone

[37] *Applications of the theory of Boolean rings to general topology*, Trans. Amer. Math. Soc. 41 (1937), 375-481.

Robert M. Solovay

[70] *A model of set-theory in which every set of reals is Lebesgue-measurable*, Annals of Math. (2) 92 (1970), 1-56.

Franklin D. Tall

[74] P-*points in* $\beta N \backslash N$, *normal non-metrizable Moore spaces, and other problems of Hausdorff*. In: TOPO 72 - General Topology and its Applications, Lecture Notes in Mathematics, #378, pp. 501-512. Springer-Verlag, Heidelberg, 1974.

[*] *An alternative to the continuum hypothesis and its uses in general topology*, manuscript.

Hisahiro Tamano

[60] *On paracompactness*, Pacific J. Math. 10 (1960), 1043-1047.

[62] *On compactifications*, J. Math. Kyoto Univ. 1 (1962), 162-193.

Alfred Tarski

[28] *Sur la décomposition des ensembles en sous-ensembles presque disjoints,* Fund. Math. 12 (1928), 188-205.

[38] *Drei Überdeckungssätze der allgemeinen Mengenlehre,* Fund. Math. 30 (1938), 132-155.

A. Tychonoff

[30] *Über die topologische Erweiterung von Räumen,* Math. Annalen 102 (1930), 544-561.

Stanislaw Ulam

[30] *Zur Masstheorie in der allgemeinen Mengenlehre,* Fund. Math. 16 (1930), 140-150.

Paul Urysohn

[25] *Zum Metrizationsproblem,* Math. Annalen 94 (1925), 309-315.

[27] *Sur un espece métrique universel,* Bull. Sci. Math. 2me Série 51 (1927), 43-64.

I. A. Vaĭnšteĭn

[47] *On closed mappings of metric spaces,* Doklady Akad. Nauk SSSR 57 (1947), 319-321 [in Russian].

[52] *On closed mappings,* Moskov. Gos. Univ. Uch. Zap. 155 (1952), 3-53 [in Russian].

André Weil

[37] Sur les Espaces à Structure Uniforme et sur la Topologie Générale. Actualités Scientifiques et Industrielles No. 551. Hermann, Paris, 1937.

G. T. Whyburn

[50] Open Mappings on Locally Compact Spaces. Memoires Series No. 1. American Mathematical Society, New York, 1950.

[52] *On quasi-compact mappings,* Duke Math. J. 19 (1952), 445-446.

Raymond Louis Wilder

[49] Topology of Manifolds. Colloquium Publications

Volume 32. American Mathematical Society, Providence, R.I., 1949.

S. Willard

[66] *Absolute Borel sets in their Stone-Čech compactifications,* Fund. Math. 58 (1966), 323-333.

D. E. Wulbert

[69] *A characterization of* C(X) *for locally connected* X, Proc. Amer. Math. Soc. 21 (1969), 269-272.

Index of Symbols and Terms

Special Symbols